ESSAYS ON

*William Chambers Coker*

PLATE 3

Fig. 1. Boletus fraternus.  No. 12811.
Figs. 2, 3. B. rubinellus.  No. 3606.
Fig. 4. B. indecisus.  No. 9869.
Fig. 5. B. luridus.  No. 3329.  Note: Stem usually more yellow.

ESSAYS ON

# *William Chambers Coker,*

# *Passionate Botanist*

MARY COKER JOSLIN

*with drawings by*
  *Sandra Brooks-Mathers*

University of North Carolina at Chapel Hill Library
Botanical Garden Foundation, Inc.
*Chapel Hill, North Carolina*

*Copyright © 2003 by Mary Coker Joslin*
*Published by the University of North Carolina at Chapel Hill Library*
*and the Botanical Garden Foundation, Inc.*

*All rights reserved*
*Manufactured in the Canada*
*Design & production by B. Williams & Associates*
*Typeset in Cycles and Arepo*

Library of Congress Cataloging-in-Publication Data
Joslin, Mary Coker, 1922 –
    Essays on William Chambers Coker, passionate botanist /
    Mary Coker Joslin ; with drawings by Sandra Brooks-Mathers.
        p. cm.
    Includes bibliographical references (p. ).
    ISBN 0-9721600-0-0 (alk. paper)
    1. Coker, William Chambers, 1872–1953.
    2. Botanists—North Carolina—Biography. I. Title.
    QK31.C65 J65 2003
    580'.92 —dc21   [B]     2002034189

First edition, first printing

*Frontispiece, left:* Portrait by Arthur Bye that hangs in William Chambers Coker Hall at Coker College in Hartsville, South Carolina. A mushroom illustrated in the plate lies beside Coker's chair. The view is of his garden from the porch at "The Rocks." Significantly depicted are his *Cedar of Lebanon* in the distance and the smilax vine growing on the porch post behind him. (He had recently completed his smilax article at the time the portrait was painted.) *Permission of Coker College. Photograph by James Jolly.*

*Frontispiece, right:* Plate 3 of Coker's book *The Boletaceae of North Carolina,* the page represented in the open book on his lap in the portrait (facing page). *Permission of the North Carolina photographic archives, Wilson Library.*

*For William Joslin*

*Smile O voluptuous cool-breath'd earth*
*Earth of the slumbering and liquid trees!*
*Earth of the departed sunset—earth of the mountains misty-topt!*
*Earth of the vitreous pour of the full moon just tinged with blue!*
*Earth of shine and dark mottling the tide of the river!*
*Earth of the limpid gray of clouds brighter and clearer for my sake!*
*Far-swooping elbow'd earth—rich apple-blossom'd earth!*
*Smile, for your lover comes.*

*Prodigal, you have given me love—therefore I to you give love!*
*O unspeakable passionate love.*

—Walt Whitman, "Song of Myself"

# Contents

# Illustrations

# Foreword

LIKE MOST institutions, universities have eras of greatness that catapult them into worldwide recognition and thrust them into roles of scholarship and intellectual and cultural leadership not previously experienced. No such advance occurs, however, without strong, productive, and creative faculty members teaching eager students, engaging in superior research and advanced study, and extending themselves into related public service.

William Coker was one of that sterling group of faculty colleagues who, by virtue of superior scholarship, brought the University of North Carolina into this national and international community of learning. Receiving his Ph.D. in 1901, he brought honor to the Johns Hopkins University by his dissertation on the developing seed of the bald cypress, a work that became in 1903 the first publication from the Botanical Laboratory of the Johns Hopkins University. William Coker came to Chapel Hill in the fall of 1902, and for the next forty-three years he fulfilled the role of distinguished scientist and scholar. Recognizing his valuable service, the University of North Carolina awarded him a Kenan Research Professorship in 1920.

Dr. Coker was an exceptional teacher, always stimulating his class to original inquiry, and, over the years, he produced an impressive galaxy of professional scientists. His published work on aquatic and fleshy fungi stimulated worldwide inquiry and activity that gained him early recognition as an innovative scientist and scholar.

It was Dr. Coker's gift of a half-century of individual attention and care for the natural beauty of the University campus, exemplified by the establishment and development of the Coker Arboretum, that inspires even the casual visitor to a new and richer appreciation of plants and trees and our relationship to all things natural. For those of us who love and deeply appreciate the natural beauty around us, our debt to him is great. We thank Mary Joslin for her enormous scholarship and uncommon devotion to bringing this remarkable story to us.

Dr. Coker's legacy is this most beautiful campus, a place of restful meditation and great natural charm that renews our spirit and sense of well-being. In this time of hurried existence, let us heed the lessons exemplified by the life of this distinguished scholar, teacher, and good public servant. We will be better and much wiser citizens when we do.

*Chapel Hill*　　　　　　　　　　*William C. Friday*
*Fall 2001*　　　　　　　　　　President Emeritus
　　　　　　　　　　　　　　University of North Carolina

# *Preface*

AFFECTION AND personal admiration are part of my irresistible impulse to write these essays about William Chambers Coker, my uncle. My earliest memories include him. During walks in the garden of our parents in Hartsville, South Carolina, Uncle Will would question us children about what we were seeing. "Children, how do you know an oak tree from other trees?" he would ask. Knowing his young companions incapable of giving anything like a reasonable response, and liking to be the teacher, he would supply a simple answer: "By its acorns." His interest was not confined to flora. He was alert to observe and to communicate to us information about any denizen of the rich natural world. Once, after dinner on a hot summer night, he called my attention to a small insect meandering across the tablecloth and to the fact that the odds were that this particular creature had never been studied or classified. When our elders remarked quick wit and keen interest in nature in a member of the younger generation, one might hear the whispered comment, "We just may have another Will."

A few curious childhood observations about Uncle Will remain with me. For example, he had a curved little finger on his left hand, obviously a birth trait, as it is clearly visible in the childhood photograph of him with his brother David, my father. Along with my sister Carolyn, twelve months younger than myself, I was once struck by a curious phenomenon one could not ignore. Uncle Will kept a long, hard, and sharp fingernail on his smallest right hand finger. Once when he saw us staring at it rather rudely, he explained that this was convenient for him in examining certain seeds.

In addition to sight, his other senses served him well in his observation of nature. The sound of a bird alerted him to nearby plants. Smell, touch, and taste were all used. An unmistakable scent called attention to a woodland magnolia not yet in sight or to wax myrtles where one could expect to see the myrtle warbler feeding on its berries in winter. A slight pubescence on the underside of a leaf or fuzz on an acorn could determine the species of oak. Uncle Will encouraged others to identify by sight, scent, and sound. He said in his description of the parsley haw,

*Crataegus marshallii*, his favorite plant in UNC's Arboretum, "In spring, it is covered with a cloud of small flowers which are very fragrant and full of bees." He would often taste a leaf, a piece of bark, or a seed to test a plant's identity, or simply to enjoy its properties.

Uncle Will died when I was twenty-eight years of age. We had taken our second child to see him the previous spring. I have vivid memories of him, from my earliest childhood to graduate school at Chapel Hill in the mid-1940s and thereafter. The memories of others enrich my own. I was able to take advantage of the closing window of time when I could still record the experiences of students, relatives, and friends who knew him in person. Though these recollections help bring the subject alive for me, and I hope for the reader, they are necessarily filtered through many years of experience. It is the study of the personal papers and published writings of William Chambers Coker that provide the more reliable substance of the remarks that follow. Through them, I have come to know him and his work from a different perspective and in greater depth than previously.

It seems appropriate to celebrate the life of the passionate botanist in conjunction with the centennial in 2003 of the Coker Arboretum, the garden on the central campus at Chapel Hill that he founded, planted, and tended for fifty years. At the present moment, when we are confronting the appalling loss of much of the floral opulence of the Carolinas and of our national and worldwide plant resources, it is timely to remember W. C. Coker's research in botany, his field studies, his contribution to our landscapes and gardens, and his passionate efforts to understand and preserve our precious areas of environmental value. His life work sounds a tocsin poignantly reminding us of the current crisis of our diminishing natural heritage. Above all, however, we remember his inestimable joy in the study of nature in all of its manifestations.

A word is in order about the formal aspects of this work. In the appended lists of Coker's publications and the plants named in his honor, I use the scientific style found in the Couch and Matthews article in *Mycologia* (1954). This style is followed of course in the research of William Burk, who compiled for me the list of plants named in honor of Coker. Otherwise, for the list of works cited in this study and for text notes I follow the Modern Language Association style, with which I have become familiar. I have italicized all botanical names of plants except in the case of plants whose scientific names are used also as the common names. In quoting letters and other informal documents, I have in rare instances made small alterations in spelling, style, or punctuation for the sake of clarity.

Abbreviations include SHC to denote the W. C. Coker papers, #3220 in the Southern Historical Collection, housed in Wilson Library at the University of North Carolina at Chapel Hill. The dates and names of the correspondents mentioned in the text itself will usually be a sufficient guide to the relevant folders in the W. C. Coker papers. *JEMSS* refers to the *Journal of the Elisha Mitchell Scientific Society* and SCL to the South Caroliniana Library. UNC refers to what today is officially named the University of North Carolina at Chapel Hill, or UNC–Chapel Hill; yet the old short abbreviation, UNC, remains in wide use. William Chambers Coker is usually referred to as Coker, though Professor Coker, Dr. Coker, and Will are sometimes used, according to context. Brackets are used for clarity or to indicate an uncertain reading in manuscript.

# Acknowledgments

TO ALL OF THE PERSONS who have helped me in this study, too many to name, I give my hearty thanks. Certain individuals, however, must be mentioned specifically for their generous help. From the beginning, Peter White, Charlotte Jones-Roe, Ken Moore, James Ward, Sandra Brooks-Mathers, and other members of the North Carolina Botanical Garden staff encouraged me in this enterprise. I wish most particularly to call attention to Sandra's creative work in designing the appropriate small graphics which introduce each chapter of this book. The enthusiasm of C. Ritchie Bell, former director of the Garden, and James and Florence Peacock, who live on land formerly included in Coker's garden at "The Rocks," helped convince me of the value of this project. The belief of these friends that a close examination of the life and work of William Chambers Coker would be an appropriate contribution to the centennial celebration of the Coker Arboretum at UNC–Chapel Hill launched me in this task and helped me bring it to completion.

To William Burk, himself a mycologist as well as librarian of the John N. Couch Biological Library in Coker Hall at UNC–Chapel Hill, I express my thanks for his essential help in locating sources, guiding me through mycological mysteries, and listing for this project the fungi and higher plants named in honor of W. C. Coker. He supplied me with many reprints of Coker's writings in his care and led me to other relevant books and articles. I also thank Rogers McVaugh, learned botanist at Coker Hall, who kindly gave me the benefit of his knowledge.

Joe Hewitt, provost for libraries at UNC, was a great source of strength to me in his belief in the importance of this project and in his help in my bringing it to fruition. Richard Shrader, John White, Jill Snider, Rachel Canada, and others at the Manuscripts Collection in Wilson Library granted me access to the William Chambers Coker papers and other pertinent materials in the archives of the University. Alice Cotten helped me with papers in the North Carolina Collection, and Neil Fulghum kindly offered me access to a rare book in the exhibition gallery there. I wish to thank also my friends, Michele Fletcher and Marcella Grendler, for permitting me access to Coker papers through photocopies during a period when I could not work at Chapel Hill. Jerry Cotten and Keith Longiotti without complaint fulfilled my repeated orders for archival photos and other pictures related to this project.

I thank Jennifer Allain Rallo of the Ferdinand Hamburger Archives of the Milton S. Eisenhower Library at the Johns Hopkins University for sending me photocopies of documents related to W. C. Coker and his graduate studies at Johns Hopkins. The staff of the South Caroliniana Library on the campus of the University of South Carolina allowed me to study correspondence between William Chambers Coker and his family. Samuel Ford was most helpful in guiding me to relevant papers there. Minou Monakes of the J. L. Coker III Library at Coker College took time from her work to locate materials there on William Chambers Coker, including papers concerning his studies at Bonn and an unpublished photograph of Coker as a young man. James Daniels, president of Coker College in Hartsville, South Carolina, sent me a color photograph of the portrait of W. C. Coker that hangs at the entrance of the William Chambers Coker Science Building in Hartsville. My thanks go to him and also to the Coker College photographer, James Jolly. Vickie Eaddy of the administration office at Coker College supplied me with information on Dr. Velma Matthews. I am grateful for the help and advice of Albert E. Sanders, curator of natural sciences at the Charleston Museum, and Sharon Bennett, archivist of the Charleston Museum. Regina Oliver and Tracy Chrismon of the UNC–Chapel Hill General Alumni Association helped me to locate information in their care on the students of W. C. Coker.

My hearty thanks go to those persons who have read and commented upon early drafts of this work. They include Susanne Gay Linville, Margaret Watson Cooper, Martha Coker Ziegler Huntley, and Preston Wescott Fox, nieces of William Chambers Coker and Mrs. Louise Venable Coker. Edwin Linville, James Coker Fort, Jean S. Fort, John E. Lee, Ione Coker Lee, Charles W. Coker, Joan S. Coker, Thomas Stanback, Margaret L. Stanback, J. Thomas Rogers, and Florence L. Snider, all relatives of W. C. Coker, carefully read my manuscript and gave me the benefit of their comments. Other good friends of the University of North Carolina, some of whom knew W. C. Coker, read drafts of certain chapters of this work. These include William C. Friday, C. D. Spangler, C. Ritchie Bell, Betsy Sawyer, Helen Kelman, Claire Freeman, T. Franklin Williams, Carter C. Williams, Mary Beth Coker Britt, and George Sawyer. Albert Radford, Laurie Stewart Radford, and Jim Massey helped me to understand the history of the University of North Carolina's Herbarium. C. Ritchie Bell and Edgar H. Lawton Jr. read and approved a late version of the text. Those who helped me search for pertinent photographs include Charles S. and Erd Venable and staff members of the Hartsville Museum, as well as members of the Coker family.

David Perry was generous in responding to my requests for advice. Ginger Travis, with her knowledge of and enthusiasm for the natural world and her literary skills, was just the right person to help me put these essays in good shape. I am grateful for her keen interest in this project and for her expert help. My thanks go also to Judith Panitch, Joan Ferguson and Barbara Williams for their advice on the technical aspects of preparing this project for publication. I thank Frances Hargraves who helped me in my effort to identify some of the gardeners at the Arboretum and on the UNC campus grounds.

The following persons, as students of Coker, Couch, Radford, and Bell, enriched these writings with their reminiscences: C. Ritchie Bell, Albert E. Radford, Laurie Stewart Radford (also in her role as historian of the Herbarium at UNC), Paul Titman, Helen Parker Kelman, Claire Freeman, and my husband, William Joslin.

Two of W. C. Coker's doctoral students, John N. Couch and Velma Matthews, published in *Mycologia* in 1954 a tribute to their mentor that includes an exhaustive list of Coker's publications. This fine article provided me new insights into his life, his work, and his personality through the eyes of two former students who knew him well. The list of Coker's publications here appended is largely their careful work.

Robert Wyatt, executive director of the Highlands Biological Station, and his predecessor, Richard C. Bruce, gave me a better understanding of W. C. Coker's role there. Robert Wyatt, a botanist who studied at Chapel Hill with two of Coker's own students, guided my husband and me through the Highlands area and the Station's campus and kindly made many helpful comments upon early drafts of my essay on the Highlands Station. Florence Inman of Atlanta, present owner of the Coker Cottage across the lake from the laboratory at Highlands, kindly allowed us to photograph her home. She later encouraged me by expressing interest in my efforts through correspondence.

Sally Couch Vilas, daughter of John N. Couch, provided me with helpful information. I wish also to thank our daughter, Nell Joslin Medlin, and our friend, William Davis Snider, both writers and editors, for their kind help with my writing style. Will Joslin, our son, was always quick to guide me through the mysteries of the computer.

My husband, William Joslin, was a student of Dr. Coker's and later came to know him well as his wife's uncle. Sharing my enthusiasm for the subject, he criticized my work for both accuracy and stylistic expression. He researched deeds of Coker's land transactions and the provisions of his will in the courthouses of Macon County and Orange County, North Carolina, and Darlington County, South Carolina. His skills as a lawyer and writer, his wise advice, and his constant encouragement have been indispensable to me in this work. This book is lovingly dedicated to him.

# The Personality of William Chambers Coker

WILLIAM CHAMBERS COKER, the first professor of botany at the University of North Carolina at Chapel Hill, served the University from his arrival in 1902 until his retirement in 1944. Thereafter, as research professor emeritus, he continued to work and publish until shortly before his death in June of 1953 at the age of eighty. His contribution to the University extended beyond teaching and research. Indeed, to tell of his life and accomplishments is also to tell of the University's turning outward to serve the people of North Carolina. These essays are simply an introduction to the story of his multifaceted life. The writer trusts that others interested in the history of the University of North Carolina in the twentieth century, in botanical studies, in our natural fields and forests, in conservation, and in practical landscape planning will continue to study Coker's contribution to botanical science, to North Carolina, and to the South.

Thanks to the preservation of much of his correspondence and to the abundance of his published works, a wealth of research material on the life of William Chambers Coker survives, providing no excuse to shrink from this project. Indeed, the size of this corpus has made necessary for the purposes of this study a drastic selection from among available materials.* Examined most closely here is a limited period of Coker's career

---

* Coker published 137 books and articles. These are preserved in the John N. Couch Biology Library, botany section, in Coker Hall at the University of North Carolina at Chapel Hill. Over 10,000 of his letters and personal papers are accessible in the Southern Historical Collection in Wilson Library at UNC. There are 276 manuscripts, mostly letters,

during his prime, the years 1919 and 1920, his forty-seventh and forty-eighth years. During this period, his research on fungi, the interest of a lifetime, was particularly intense. Especially noteworthy also are earlier and later activities that shed light on his background and character. These projects include his founding and long-term directorship of the University's Arboretum, his organization and expansion of the Herbarium of the University of North Carolina, and his role in the establishment and the ongoing work of the Highlands Biological Laboratory in the Blue Ridge Mountains at Highlands, North Carolina. Not to be ignored is his role in the establishment of Kalmia Gardens in Hartsville, South Carolina, and in the plantings at Brookgreen Gardens, near Pawley's Island, South Carolina. Also considered are his lifelong moral support of his family and students and his contributions to Coker College, where he early designed the campus and laid out its planting and where the science building now bears his name.

A passion for botany was William Coker's vital force. Nothing he observed in the plant kingdom was beyond his concern. His decades of uninterrupted research dealt with widely diverse areas of botany. His dissertation at Johns Hopkins, his postdoctoral work at Bonn, and some of his early research dealt with seed development in conifers. Early in his career he was internationally recognized for his publications on water molds and fleshy fungi. His interest in these forms of plant life never flagged. In addition, he studied and published on herbaceous plants, vines, native shrubs, and the trees of the South. He considered the accurate recording of habitat range for the native plants of the Carolinas of paramount importance.

True to his upbringing, Coker was a child of the Jeffersonian Enlightenment in that he believed in the capital importance for society of the educated individual. He early expressed, in his presidential address before the North Carolina Academy of Science in 1910, his view that the inner motivation of the student, supported by his earnestness and tenacity, is the key to learning. Training alone is insufficient. He stressed education that teaches thoroughness and is practical and useful in the situation in which the individual lives. Such an education, he believed, would elevate the spirit. He expressed the hope that it would bring the student "to perceive the majestic pageantry of things seen, the vast creative power of things unseen."[1]

in the South Caroliniana Library at the University of South Carolina in Columbia. A small collection of W. C. Coker papers are in the archives of Coker College in Hartsville, South Carolina.

Coker patiently played a role in the development of his students. His joy in their tenacious work on a problem of mutual interest was infectious. By working beside his graduate students in the laboratory and by including them on collecting trips, he gave tangible evidence that they were trusted colleagues. He followed the progress of former students after they left the University, often having assisted them in finding professional employment. Other students he helped in tangible ways—by providing for them a room in his home or a loan for tuition to prevent their leaving the University for financial reasons. Conversations with certain of his former students reveal evidence not to be found in the written record of his genuine concern for them.

Coker's effective service to the University as director of grounds and buildings can be traced to his belief in the civilizing influence of natural beauty and to his natural talent as a landscape planner.[2] He possessed an instinct for aesthetic effect and a practical knowledge of what would succeed under specific environmental conditions. The village of Chapel Hill reflects his long-term residence there and his interest in promoting beauty for the total community. R. W. Madry, in a newspaper article with the headline "Beauty of Chapel Hill is Monument to Coker," remarked, "Dr. Coker has grown great numbers of shrubs and trees which he has given away, both to the campus and to other private owners for the beautification of the town."[3]

The Nolen plan for the central campus of the University and the area south of South Building, shown in a sketch completed in February of 1919, owes much to Coker. John Nolen, distinguished landscape planner of Cambridge, Massachusetts, who had been a student of Frederick Law Olmsted, designed in 1919 a plan for the expansion of the University that was partially implemented during the 1920s. As faculty chairman of grounds and buildings, Coker was Nolen's chief correspondent on behalf of the University before and after the deaths of President Edward Kidder Graham and of the acting president, Dean Marvin Stacy, both of whom died in quick succession during the influenza epidemic of 1918 and early 1919. Nolen incorporated some of Coker's suggestions in his final plan. After the Nolen plan was completed, Coker was instrumental in the adoption of some of its features.[4]

Coker assumed public service as a personal obligation by regularly offering his native talent as a landscape designer to individuals and groups in North and South Carolina. His practical advice was requested and generously given for the landscaping of churches and such public sites as a library, a cemetery, a railroad station, and a post office. Sweet Briar College and Coker College profited from his knowledge of landscape design. Friends and neighbors in the village of Chapel Hill and members of his family in Hartsville requested and received advice for their gardens. He served the state of North Carolina through the University's Extension Program for the landscaping of schools, the type of public service that President Edward Kidder Graham considered a responsibility of the state university. Considering himself a servant of the people as a state university faculty member, Coker traveled extensively through North Carolina to design plans for public school grounds.

That others acknowledge his work was of little concern to William Coker. No doubt this attribute was one secret of his quiet success. His close friend and colleague on the faculty, Dr. William de Berniere McNider, remarked on this character trait in a letter of June 17, 1920 to Coker, soon after his appointment as Kenan Professor of Botany: "I was glad of the honor, not because it means anything to you, because it doesn't, but because you can mean much to this Professorship."[5]

Throughout his years at the University, Dr. Coker never ceased research, writing, and publishing. He was among the incorporators and first board members of the University of North Carolina Press, founded in 1922. His book *The Saprolegniaceae* was its first publication, in 1923. The Press also published his monograph on the *Clavarias* the same year.[6] He encouraged scientific research and its publication during his long tenure as editor of the *Journal of the Elisha Mitchell Scientific Society*, from 1904 to 1945.

It is fortunate for this study that during the first half of the twentieth century most of the business of the University, as well as almost all personal and professional business, was conducted by correspondence. The survival of these written records permits an examination of Coker's considerable role in the affairs of the University.

Delight in close observation of the natural world was a conspicuous characteristic of William Chambers Coker. His awareness of nature was not restricted to plant life. In the 1940s, he had this to say about birds in the vicinity of the University at Chapel Hill:

The primitive seclusion of our woods, grown more withdrawn during the automobile and movie era which has educated our young people to avoid walking, is reflected in the presence of the wild turkey, which until recent years has nested on University property and even now is breeding just across Morgan's Creek from the Mason Farm. The pileated woodpecker still makes its home on University property and a few years ago came up into Battle Park a few hundred yards from the Campus and occasionally even into the Arboretum itself. Every year quail build their nests within the limits of the town. Almost every night in winter the mellow hoots and wild screams of the barred owl can be heard in the village and until recently a pair nested in Davie Poplar on the Campus. Also recently a bald eagle has been seen a few miles from Chapel Hill. Up to the present the number of species of birds recorded from Chapel Hill, including residents and transients, is 195.[7]

The personality of William Chambers Coker reflected his home and family. Certain characteristics of his father were his own: pleasure in the perception of events in the natural world, diligence for the task at hand, willingness to take risks, carefulness in daily affairs, and the ability to move steadily toward the fulfillment of an idea without undue discouragement at setbacks. Strangely enough, a final outing of Will's life seems to mirror that of his father. Both had to do with admiring the wonder of birds. The last time his father left his home in Hartsville, he crossed the street to the garden of his son David to see the nest of a bluebird, a rare bird for the area at the time.[8] Similarly, Dr. Paul Titman described his own last visit with his old professor, William Chambers Coker:

He [Dr. Coker] said, "Have you been through the garden lately?" Over his head Louise Coker nodded. He said, "Let me put on a jacket." He did not change his shoes, even though he still had on his bedroom slippers. While he was gone, she said, "He's not been outside the house in almost a month. I am so glad that he wants to go through the garden with you." . . . We came back. I said goodbye and started down the road to the entrance. And, halfway there, I looked up and saw a great horned owl sitting in an oak tree, in the middle of the afternoon—uncanny. I went back and rang the doorbell. They came back to look at the owl. And this is the last time I saw him, standing there, looking up with keen interest and awe at this phenomenon of nature. He died two weeks later.

During all of Coker's years in North Carolina, Hartsville remained "home" for him. As one of his students remarked, "It was interesting also to see that Hartsville was deep in his thoughts. He always compared things as to the largest, the rarest, or the best north of Hartsville or south of Hartsville. That was the meridian: Hartsville, the home of the Coker family to begin with. I never was at Hartsville and I've often wondered if it is as wondrous as he seemed to imply that it had been."[9]

It was important to Coker to be surrounded by open land, the environment of his early years. In 1906 he bought farmland on the outskirts of Chapel Hill.[10] For a time, he devoted part of this land to farming. In 1944, the Coker estate was described as "seventy-five to eighty acres with a fine orchard of peaches, apples, pears and grapes, freely pilfered by neighborhood children."[11] At the time of his death in 1953, fifty acres still surrounded his home.

Coker was reserved about his own activities. At the time of his retirement in 1944, a newspaper writer called him "as shy an individual as you'll find at the University," one who has "always shunned the limelight."[12] Though abrupt in correcting what he perceived to be errors of observation or lack of diligence on the part of a student, he was quick to encourage good work. A niece who made her home with the Cokers from time to time in their later years observed that he could smolder silently when upset or angry but did not rant. This silence was a signal for her and her aunt to disappear.[13]

The society in which Coker had been raised was much more homogeneous than that which we now know. The dominance of the patriarchal white male was rarely challenged during the span of Coker's active life. Some of his comments on minorities and women during the early decades of the twentieth century witness to this general cultural attitude. In describing the business career of his father, William characterized a Jewish business associate as shrewd but "essentially honest."[14] In quoting black friends and associates, he paid tribute to their folk wisdom but considered them in need of protection at a time when blacks had made little progress in the mainstream of southern life and were in jeopardy in many areas of North and South Carolina.[15] He respected "ladies," treating them with the manners of a courtly gentleman. Though giving every opportunity to one of his most brilliant Ph.D. students, the feminine Velma Matthews, later professor of biology at Coker College, he was protective of the women in his family. He discouraged his niece Dorothy Coker, a graduate of Wellesley College and illustrator of mushrooms in some of his publications, from applying to teach in his department where there were "large classes of young men, some of them tending to be unruly."[16] Coker depended on Miss Alma Holland, his able assistant and coworker for many years, to teach botany classes. During these sessions, however, he generally sat at the rear of the classroom to lend her authoritative support and to discourage any possible uncouth behavior. Coker readily sought essential information from women who were botanists, women such as Dr. Gertrude Burlingham, Miss Ann Hibbard, Miss Alma Holland (later, Mrs. Beers), Mrs. Ida Jervey, Mrs. Frances Harper, and Miss Alice C. Atwood. Presumably, women such as these, given the opportunities afforded women today, could have become university professors like himself.[17]

While sharing the attitudes of a typical southern gentleman of his time, Coker also was a close friend and supporter of the two idealistic presidents of his University: Edward Kidder Graham (1914–18), whose belief that the University existed for the service of its state Dr. Coker shared; and Frank Porter Graham (1930–49), cousin of E. K. Graham, who was ahead of his time in championing minorities and other power-less persons in need. Coker was disturbed by the racist tactics used to defeat the second President Graham in his campaign for United States Senate in 1950. He contributed generously to Graham's campaign more than once.[18]

There is evidence of Coker's keen interest in the affairs of the nation and in issues that affected the state of North Carolina and the village of Chapel Hill. He was troubled by the reluctance of the United States to give greater support to France and England during the terrible year of 1915. A draft in Coker's hand of a telegram to President Woodrow Wilson, with the notation "sent May 11, 1915," urged a "firm and final protest by all neutral nations against the cumulative inhumanities of the Teutonic Allies." Twenty-six of Coker's colleagues on the faculty of the University joined him in signing this communication.[19] Even more disturbed by the continued neutrality of the United States in the face of the slaughter in Europe, Coker wrote an eloquent "open letter" to President Wilson, dated December 20, 1915, voicing a plea that this nation "be a constructive force for good in these sad times." He characterized the United States "not as an inactive supporter of rights or a schemer for greater prosperity, but as the protagonist of an idea." He further asked that the President announce the intention to prepare for war "against those forces of despotism which stand against the progress of liberty."[20] On April 2, 1920, in separate letters to two French mycologists, Léon Dufour and Narcisse Patouillard, Coker apologized for the Senate's peace treaty, which he considered "a betrayal of the interests of the American people."[21]

The anti-evolution bills of 1925 and 1927 introduced into the legislature of North Carolina by Representative David Scott Poole, with their threat to academic freedom, concerned William Coker greatly as a citizen, a professor, and a leader in the North Carolina Academy of Science. At the twenty-third annual meeting of the Academy in December of 1924, the Resolutions Committee, headed by Dr. Bertram Whittier Wells of North Carolina State College, presented an answer to critics who dismissed evolution as a discarded and discredited theory.[22] With the possibility of an even more stringent limitation to academic freedom by a second Poole bill to be introduced in January to the 1927 session of the North Carolina legislature, the Academy passed a resolution, proposed earlier in the day by B. W. Wells, which strongly supported freedom of all teachers to discuss any matter of scientific concern. W. C. Coker was chairman of the committee presenting this resolution to the Academy that endorsed "most emphatically the stand of Dr. H. W. Chase, president of UNC, and Dr. W. L. Poteat, president of Wake Forest, on the freedom of thought and teaching." The resolution passed unanimously by a standing vote of the membership in attendance at the session.[23]

Coker was incensed to learn that his former student J. V. Harvey, professor at Oklahoma Baptist University, had been dismissed for teaching evolution. Writing to Harvey on April 4, 1927 to say he was "surprised and disgusted at the news," Coker mentioned that he [Harvey] was to work at Chapel Hill the following year. Then, he added, "At this rate, I suppose your college will soon have its staff composed entirely of preachers, all teaching the same subject under different names! That would be an easy solution of all their difficulties in avoiding the search for truth."[24]

The city fathers were well aware of Coker's interest in embellishing the village of Chapel Hill. In 1916 he was appointed to fulfill an unexpired term on the Chapel Hill Board of Aldermen.[25] He served on the Chapel Hill tree committee, and, when consulted, he advised the town on landscaping the downtown area.

Coker's financial independence from his University salary during most of his career, resulting from early investments in Chapel Hill real estate and interest in several family enterprises in Hartsville, may have emboldened him to express himself on social and political issues and to launch projects of his own. One might cite as examples of such projects his own research, campus beautification, the Highlands Biological Station, the acquisition of a fine botanical library for the University, and the assistance of individual students. But Coker chose to live simply, spending much of his time in laboratory work or out-of-doors as a student of plants in their natural environment. His travels were always closely linked with botanical studies. Wherever he went, he collected specimens or visited herbaria.

Coker was well-read and open to ideas developing in the sciences. He latched onto new ideas. In response to a letter of January 1925 from Dr. J. A. McKay, who had sent him a copy of the *New Republic* with an article on sex transformation in animals, Coker commented in an almost prophetic tone, "The new discoveries in physiology are upsetting to all our ideas. If we can hold on to a stable society for another two hundred years, wonderful things may take place in the modification of human life and thought."[26]

He could, however, make authoritative statements that he later might have wished to modify. For example, after Dr. B. W. Wells of North Carolina State College presented a paper on the effect of salt spray in the destruction of seaside vegetation to the 1938 meeting of the North Carolina Academy of Science, Coker is reported firmly to have expressed his belief that the wind alone was responsible for such destruction.[27] The recent writings of Orrin H. Pilkey and others on the North Carolina barrier islands assume the accuracy of B. W. Wells's observations. Pilkey and his coauthors use the expression "salt-spray-pruned vegetation" without the need to explain its meaning.[28]

Though a bachelor during the greater part of his career at Chapel Hill, and though usually reserved in public, Coker was sociable with family and friends and enjoyed being a host. He entertained in his pleasant home and garden for many years without a hostess. On the back of a letter from his brother David dated December 10, 1918, he jotted down a menu and a guest list of twenty, composed chiefly of faculty friends.[29] At the age of sixty-two he married a lifetime friend thirteen years his junior, Louise Venable, daughter of the University president who had hired him. She shared his friends and his interests and was a gracious hostess at "The Rocks" for many years.

A serious, tenacious worker during the day, Coker was often convivial and not averse to spirituous drinks with friends in the evenings. He referred to the wastebasket beside his library chair as the "VIB," the very important basket where his highball could be quickly hidden from the unexpected entry of teetotalers among his friends and relatives. In the opinion of one of his students, Mrs. Coker as hostess at "The Rocks" was an influence for moderation. In reply to a question about whether or not his fondness for a drink affected his work, several former students answered firmly in the negative.[30]

William Coker regularly invited to his home family members, visiting speakers, or traveling botanists whom he had come to know by their publications and by correspondence.[31] In April of 1918, Dr. Charles Sprague Sargent of the Arnold Arboretum in Boston visited him. Coker revered Sargent as "the greatest American authority on trees."[32] Dr. J. K. Small of the New York Botanical Garden visited Coker in 1920 and took photographs of his garden. From time to time his guest room was made available to nephews, nieces, children of cousins, and especially to students in need of a boost.

Though Coker was neatly and appropriately dressed, his wardrobe was spare. When asked to assume the role of father in the afternoon wedding of one of his nieces in Hartsville in 1946, he and Mrs. Coker were in a flurry over his "giving-away" suit. According to another niece who was with them at the time, he ended up borrowing an appropriate suit from a friend. The same niece described his clothing at the time: "By choice he had a very old and limited wardrobe, two light weight gray suits, an old shapeless tweed jacket, a few slacks, a worsted coat that he had purchased on their honeymoon, (twelve years before), leather high-top shoes."[33]

Coker's whimsical sense of humor smoothed his relations with some-times prickly colleagues. His training in courtesy generally set the tone for his letters. Barbed criticisms of Coker by the irascible Lloyd of the myco-logical registry in Cincinnati evoked a light comment, rather than a reply in kind.[34] He declined pessimism in reporting the near-tragic loss of a con-siderable portion of a month's collecting foray with three colleagues in 1922 in the North Carolina mountains. By using understatement and a hu-morous turn, he reconciled himself to this largely irreparable loss by de-scribing it as the disappointing booty of a misguided thief who fancied he had something of value.* He did not reject the dubious claims of amateur botanists without testing them first. Upon hearing that a gentleman had seen the carnivorous venus flytrap in the area of Savannah, which was too far south to expect its occurrence, Coker took the matter seriously enough to make a trip to the Georgia coastal town to join the claimant in a fruit-less field investigation before mildly attributing the report to fantasy. He later commented on their lack of success with his characteristic dry hu-mor, "He was not able to find it again, if indeed he ever had found it."[35]

Dr. Coker was very fond of dogs. As he turned seventy, he replied to a request for the philosophical thoughts of a patriarch with this bit of ad-vice, "Marry the right woman and arrange to always have around a con-genial dog."[36] His childhood pet "Scott" eagerly accompanied him, his father, and brothers on Sunday afternoon nature walks.[37] Some of the cherished dogs of his maturity were "Tim," "Mickey," and "Tinkerbell." A dinner guest could observe him surreptitiously handing under the table a tidbit to his current canine companion when Mrs. Coker's head was turned. Mickey, Tinkerbell, and Mrs. Coker's dog "Sinner" were buried in the dog cemetery beyond their lawn.[38] He was fond of "Dickie," the pet of his brother David's younger daughters. When this little dog of mixed breed was injured by a car at Myrtle Beach during a visit with his brother's

* "During the past August the author with Mr. H. R. Totten, Miss Alma Holland and Mr. J. N. Couch spent two weeks in Blowing Rock and the neighboring mountains getting together a valuable lot of fungi, many of them *Clavaria*. These are the ones referred to as 'collected by Coker and Party.' About one-fourth of the material (with notes) prepared on this trip was unfortunately lost on the return, being mistaken for something valuable by a misguided thief." William Chambers Coker, *The Clavarias of the United States and Canada* (Chapel Hill: U North Carolina P, 1923) 3–4.

family, their Uncle Will picked up the yelping pet and drove him to a veterinarian, had his leg set, and brought him back, no longer in pain, merrily stumping along with his leg braced by a splint, much to the relief and gratitude of his nieces.

Laurie Stewart Radford recounts a misadventure with a happy ending which occurred during her stay at Highlands in the Coker Cottage on Lake Ravenel in the summer of 1936, where she assisted Dr. Coker in the collecting, recording, and drying of mushrooms.

> Dr. Coker's constant companion day and night was a little one-man dog named Mickey (and I mean "one-man") who looked just like the little dog of "His Master's Voice," the famous Victrola trademark. Many mornings the Cokers, Mickey, and I walked the pleasant trails around Highlands in search of mushrooms and other plants for study. On longer trips by car, sometimes over a day or two, Mickey of course always went along. The Cokers sat up front and took turns driving while I sat in the back and made certain the little creature was on leash before a door was opened. I still keenly remember the day I fell from grace. . . . We had just returned to Highlands after a wonderful day of collecting mushrooms. The Cokers stopped as usual at the post office to get their mail. Mrs. Coker opened the front door and before I could get hold of Mickey he sprang over the front seat, shot through the partially opened door, and disappeared downtown. To me it seemed like a long afternoon at the little cottage, without much conversation. Finally, the little rascal walked in without apology, hungry and tired and ready for supper. He ate, then lay down at his master's feet for a long nap, full of dreams, no doubt about his cleverness in escaping from me for a rare afternoon of forbidden freedom and adventure.[39]

In reply to a letter from a niece offering him a puppy whom she called a half brother of Mickey, Coker accepted the gift. He wrote to her, "Mickey is a highly intelligent little fellow, and devoted to me. He is the only dog I have ever owned that would really bite. He has a peculiar complex not to allow anybody in my bedroom. My man, Rhodes, brings me water to drink every morning early but Mickey has got so he will not allow him to come into the room to hand it to me. We took him up to Highlands with us and he was quite a lot of entertainment, though he got car-sick on the way up."[40]

Coker was much affected by the death of Mickey. He revealed his grief in letters to friends in June of 1943. In expressing his regret to Ralph M. Sargent that he would not be in Highlands that summer, he wrote, "We will miss you all the more because our little dog Mickey died about two weeks ago and will not be with us for the first time in many years."[41]

Coker's love of reading, especially of articles on biological subjects and poetry, was obvious to visitors at "The Rocks" or even in his office on campus. Paul Titman, his student of the 1930s and early '40s, recorded an anecdote concerning the wife of a little-known Victorian poet who was visiting in Chapel Hill and whom he introduced to Dr. Coker at Davie Hall. The visitor was impressed that Coker had a volume of her late husband's poems in his office and was able to quote from some of them.[42] He received callers in his library. There he sat beside a bookcase with his current reading. At any moment he would lean over and select a book perhaps of poetry, perhaps on manatees, or perhaps a number of the British journal *Nature* in order to call his guest's attention to an excerpt that had recently impressed him. A resident niece remembered some of the contents of this bookcase by his library chair: his current favorite poetry, the *Elisha Mitchell Journals,* books by his friends, his tree book, Walt Whitman's *Leaves of Grass,* Charles Venable's algebra book.[43] Eventually, many of his scientific books and papers went to the Couch Library in Coker Hall. Some of the finest of his collection are now preserved in the Rare Book Collection in Wilson Library.[44] He mentioned in his will that his nonscientific books were to go to the Coker College Library.[45]

Dr. Coker could compose nonsense jingles at will. He easily produced these for family occasions. He was especially fond of English nursery rhymes. Were he a visitor in one's home, one would not be surprised to know he was awake by hearing the following or something similar:

> One misty, moisty morning
> When cloudy was the weather,
> I chanced to meet an old man
> Clothed all in leather.

Apparently early-morning nonsense rhymes were a part of his childhood. Coker wrote the following reminiscence of his father, "It was rather habitual to him to sing snatches of songs, some of his own make-up while dressing. I remember one morning as he came out of his door at breakfast time, he called to us boys upstairs:

> Breakfast bell,
> I think I smell
> That old ham bone
> I know so well."[46]

In response to a solemn event, Coker's mood was different. He jotted down ardent words on the back of a letter from his brother James which outlined the planned program for the memorial exercises to be held at Coker College some months after his father's death in 1918. One assumes this verse to be of his own composition, though it could possibly be the work of another cited from his memory's store. The lines, though not completely legible, seem to read as follows:

> Send I my astral self
> Tender and mystical
> Shed of its grosser stuff
> All that's sophistical
> Which used a candle flame
> Steadied and pointed
> As by the breath of God
> Guided and haunted.[47]

Coker's humor was whimsical. He chuckled and grinned, but seldom laughed aloud. Louis Graves, the long-time editor of the *Chapel Hill Weekly*, understood his wry humor. In an article that appeared in the *Baltimore Evening Sun* on July 10, 1946, the writer is reacting to a romantic description of the honeysuckle vine that appeared in the *New York Times*, a paean to its "sweet display when night creeps over the valley and a new moon hangs in the western sky." The *Baltimore Sun*'s writer then quotes Mr. Louis Graves, who had interviewed Dr. Coker about this supposedly charming late-spring bloomer. Graves cited Coker's description of honeysuckle as a "first class pest." According to Coker, "it runs wild in all directions, throttling and stunting trees and rendering fields unfit for cultivation." Coker calls Japanese honeysuckle, said Graves, "the worst pest this country has had since it was struck by the chestnut blight." The end of this article from the *Chapel Hill Weekly* well reflects the Graves-Coker humor. Graves wrote, "You can see the picture for yourself—the creeper, affecting an innocent and rustic air, exuding perfume, reaching out a delicate tendril toward a wild flower, a garden, or even a grove of trees, then, suddenly, clutch! A short horrible struggle and the deadly work (Mr. Coker's phrase) is done."

With some insight into the personality of the passionate botanist, we can better judge how his early life and education helped set the course he was to follow, but not without a detour along the way.

# CHAPTER ONE

# The Student, Early Life

*This is about the time of year when you and I used to go
out into the garden, about thirty-two years ago, and
plant vegetables together. The time never comes around that I
do not think how much fun it used to be.*

—W. C. Coker to his father

IN ORDER better to understand the interests and personality of William
Chambers Coker, one should know something of his family and of the
environment in which he was raised. His was a no-nonsense family, with
determination forged by the hardship of war and by the demands for
action to rebuild a disrupted society. Family members valued education.
Eagerness to learn served them well for adaptation to wrenching societal
change. Observation of the surrounding natural world was a refreshment
that constantly sustained them.

William's father, James Lide Coker, son of a successful planter and merchant in nearby Society Hill, had been a delegate to the convention called
by President Johnson during the long recess of Congress in 1865. Chosen
there to represent Darlington County in the state legislative session during the fall and winter of 1865–66, James Coker introduced a bill for the
establishment of public education, which he considered essential for the
expanded and largely uneducated electorate. Though this bill did not pass,
the educational advancement of the people of his region remained one
of Will's father's lifelong civic concerns. After Congress assumed the
responsibility for a more radical reconstruction, Johnson's attempt to re-
store government in the southern states was repudiated and a program
was instituted that gave Congress greater control.

William Coker was born in Hartsville, South Carolina in 1872 during the period of the South's recovery, a scant seven years after the end of the Civil War. Though his family provided a secure home, he spent his early childhood in a civic environment characterized by instability. Federal troops occupied South Carolina until 1877. Intimidation dominated the political atmosphere—intimidation both by Radicals, supported by federal troops, and also by members of the conservative Democratic Party, who ardently sought the restoration of white rule.

James Coker, though handicapped as a wounded Confederate veteran, began immediately in the spring of 1865, the year of surrender, to revive from ruin the farm in Hartsville in which he owned a half interest with his father and which he had cultivated since his return from Harvard in 1858.[1] He sought immediately to replant in order to provide for his family and those living on the place. Having the advantage of land and a good education, he proceeded with hope and diligence to supervise the working and harvesting of a crop of cotton and corn. Sale of the cotton in the fall of 1865 netted a small profit after payment of six laborers.[2]

A look at James's wartime experience casts light on what the previous generation of Will's family must have been up against during the late 1860s and during the decade of the 1870s, the years of Will's boyhood. Family character and loyalty had permitted Will's father to cope with adversity, even to survive. Having been struck by a miniball that shattered the bone in his thigh near the hip in the night of October 28, 1863, at Lookout Mountain, Tennessee, James telegraphed his mother, Will's grandmother, Hannah Lide Coker, to come immediately with his wife Sue and the family doctor. Susan could not accompany them, as she was soon to give birth to their son James. A strong and resourceful woman, Hannah hastily packed supplies and food. Three hours after receiving her son's summons she was traveling toward Florence, some twenty-five miles from Society Hill, to board the eleven o'clock train for Chickamauga, Tennessee. She found James in deplorable condition in the cottage where his men had laid him. At first, Confederate army doctors had scant hope for James's survival, but his youthful vigor encouraged them to give him some treatment. Hannah and James were soon captives behind the Union lines as the federal army gained the territory around their small house. Hannah Coker, acting as nurse and caretaker, helped James recover sufficiently to be moved to a prison hospital at Fort McHenry near Baltimore. From there, she traveled to Washington to visit the office of the Commissary General of Prisoners, where she found that her application for James's exchange as a prisoner had been approved. She was prepared to take her case to President Lincoln if necessary to insure her son's return home.[3]

William Chambers
Coker with his older
brother David, about
1881. *Permission of
Larry E. Nelson, from
the book* Mr. D.R.
*by Rogers and Nelson,
Coker College Press,
1994.*

After almost ten hard months, Hannah and James arrived in Florence by train at eleven at night on July 21, 1864. At the station to meet them were James's father, Caleb Coker, James's wife Sue, and several friends. The entire family had assembled after midnight at Society Hill to welcome them home.[4] Such was the staunch, loyal family that helped to forge the character of William Chambers Coker.

Children of James and Sue continued to arrive. To the two, Margaret and James, both born during the war years, were added David Robert in 1870, William Chambers in 1872, Jennie in 1876, Charles Westfield in 1879, and Susan in 1882. Will was thus the middle child of the seven who survived infancy.

James L. Coker, always called Major Coker, his military rank, recognized that if the South were ever to rise from its depressed condition, industry would have to supplement agriculture. Seeking to broaden his interests, he signed in 1874 an agreement to open a factorage and office for commission merchandise in Charleston with his friend George Norwood. The family moved to Charleston in 1878 but spent the summers in

The James Lide Coker House in Hartsville, built in 1882, where W. C. Coker lived from the age of nine or ten and where he often went to visit his family until 1922 when the house was burned. Subsequently, W. C. Coker's eldest brother, James, built another house on this property. *Permission of the Hartsville Museum, Hartsville, South Carolina.*

Hartsville, where James continued to operate his farm with the help of a manager and tenants and where he opened a general store similar to his father's in Society Hill. In 1881, the family returned to live permanently in Hartsville, where Major Coker built his large frame house in 1882 and founded a bank nearby and several industries. He, with the help of others in the area, built a railroad spur in the 1880s to this isolated, chiefly agricultural community to facilitate the movement of manufactured products from the town.[5] Recognizing a great need for education in the region, he was instrumental in the establishment in the 1890s of a high school that in 1908 would become Coker College.

Their father's example of self-discipline in pursuit of creative projects was powerful in the lives of his sons, whom he encouraged and backed financially in their respective enterprises. James, the eldest son, and Charles, the youngest, became interested in inventions, engineering, manufacturing, and corporate management. David, in charge of the store, began to work with agricultural plants, seeking by plant breeding and propagation of selected stock to develop more productive and disease-resistant seed for cotton and other agricultural crops. His father backed him in the establishment of Coker's Pedigreed Seed Company, which offered superior seed for the south of the United States and for farming areas of similar climate abroad. Will shared his father's strong interest in the natural world. Though the only son to leave Hartsville, Will remained a close family member and was consulted on all important family decisions. He regularly returned for family visits.

The passion for plants, which was to dominate the life of William Chambers Coker, took possession of him in childhood. The curious boy observed a rich variety of trees, shrubs, and flowers in the forest, swamps, and countryside around his home. Behind the Coker home near Black Creek grew ancient cypresses and white cedars, many of great size. On a high north-facing bluff above this creek, some distance west of his home, was a deposit of mountain plants rarely seen in the hot, flat cotton country of the Pee Dee region of South Carolina. Will must have walked with great delight in the richly varied woods and in "laurel land," where he observed plants such as *Kalmia latifolia*, *Rhododendron caroliniana*, and *Galax aphylla*, not usually considered Hartsville natives. Nearby, west and north of the town, was another ecological association, that of the scrub oak and long-leaf pine woods of the sandhills. Here he could find trailing arbutus (*Epigaea repens*) in spring and sandhill gentians (*Gentiana autumnalis*) in fall. He must also have observed in these early years the sandhills blazing star and a white-flowered perennial of bogs and pond margins, two native plants that were later to be named for him, *Liatris cokeri* and *Lycopus cokeri*.

Will's father James had spent the academic year 1857–58 at Harvard studying with two of the most respected American scientists of his day, Asa Gray, renowned botanist, and Louis Agassiz, Swiss-born naturalist and specialist in geology. James Coker worked in the laboratory of Professor Horsford while receiving practical instruction in botany from Gray and attending the lectures of Agassiz.* James recalled sessions at Gray's home with two other students in the class as "occasions of the greatest delight." [6] In this atmosphere, James Coker became a serious student of natural science.

Some twenty years after his studies at Harvard, Will's father was teach-

---

* David Starr Jordan says of Agassiz, "As a teacher of science he was extraordinarily skilful, certainly the ablest America has ever known. In addition he was personally devoted to his students, who were in the highest sense co-workers with him—the best friend that student ever had." Agassiz stressed study of nature out-of-doors. Jordan remarks, "The result of his instruction at Harvard was a complete revolution in natural history study in America." *Encyclopedia Britannica:* 14th ed., vol. 1 (London: 1929) 340.

Apparently David Starr Jordan spoke at Chapel Hill in 1910. Will's father wrote to ask for a copy of the talk he was to give. See letter dated January 12, 1910, in the W. C. Coker Collection, South Caroliniana Library, Columbia, South Carolina.

Photograph of a sparkleberry (*Vaccinium arboreum*) in "laurel land," now
Kalmia Gardens, taken by W. C. Coker in about 1910. *Published in Chapter III of
Plant Life of Hartsville, 1912.*

William Coker as a student. *Courtesy of the Coker family.*

ing his children by his own infectious enthusiasm for the study of nature. On regular walks in the woods and fields near their home, he called their attention to natural phenomena in all seasons. In the 1880s, he permitted Will to use the books, microscope, and considerable apparatus that he had brought back from Harvard for conducting experiments in agricultural chemistry.[7] Will wrote of nature walks with his father, "He knew quite a lot about birds, insects, and plants and would talk to us about such things, calling attention to the arrival of certain birds and their nesting habits and also to the appearance of certain insects in the spring such as katydids and locusts (cicada). He would take us to walk in the woods Sunday afternoons, and we would notice all kinds of interesting things with his help. Scott, my dog, always went with us and knew exactly when Sunday came and also the right time in the afternoon to start. If we did not start quite on time he would go up and pull Father's trousers' leg on Sunday about 4 o'clock."[8] Will was experiencing here the sort of outdoor observation of nature that his father had learned from his Harvard mentors, Gray and Agassiz. This was to remain one of Will's primary methods of study.

James Coker attempted to expose his children to nature even in the city of Charleston, where the family lived for four years, beginning when Will was five years old. In 1943, William Coker included the following reminiscence in a letter to a member of the Charleston Museum staff who had just informed him of his election as an honorary curator: "One of my earliest recollections as a boy when we lived in Charleston was being led by my father through the Museum and being tremendously awed by the buffaloes and deer. A few weeks ago I noted that the expressions on the faces of these animals had not changed at all in sixty years."[9]

Will recalled that his mother, Susan Stout Coker, daughter of a Baptist minister in Alabama, noticed his childhood experiments. One day, looking from her window, she saw him attaching a thread to a June bug. She summoned him inside to ask him how he would like to be in the place of that June bug.[10] She doubtless viewed the plant world as a morally less hazardous area for investigation than that of insects.

Two of Coker's doctoral students, in a memorial tribute published soon after his death, recorded an anecdote of his that shows that formal education, encouraged at home, was not always to the taste of the curious outdoor child. "In an unpublished manuscript, 'Childhood Recollections of My Father,' Coker relates an interesting incident of this period: 'We had at one time when I was in my early teens a . . . governess from Virginia . . . to whom I took a strong dislike which became so bad that I refused to study and did not answer her questions. This was reported to my father and he called me in and asked me what I would rather do, attend to my studies or go to work on the farm. He said that if I decided to work on the farm, to report next morning to his farm superintendent at six o'clock. I said nothing but reported for studying the next morning and behaved myself afterwards.' "[11]

On March 13, 1916, Will wrote from Chapel Hill to his father in Hartsville, "This is about the time that you and I used to go out into the garden, about thirty-two years ago, and plant vegetables together. The time never comes around that I do not think how much fun it used to be. I have been out today a good deal for the first time, taking advantage of the good weather to get some planting done in the cemetery, [and] the arboretum. . . ."[12] In response to an inquiry in 1928 about the germination of holly seed, Coker recalled a childhood experiment. "When I was a small boy," he wrote, "I planted some holly seeds (perhaps a dozen). One came up and has now grown to a very beautiful berried holly in the old home place in Hartsville."[13] This childhood interest in planting and watching the growth that followed was to remain one of Will's lifelong delights.

### Education

William Chambers Coker received the Bachelor of Science degree in 1894 as a "highly distinguished graduate" of South Carolina College, later to become the University of South Carolina. Couch and Matthews remark

William Chambers Coker
as a graduate of South
Carolina College in 1894.
*Coker family album.*

that he was on the varsity tennis team and was an active member of the Euphradian Literary Society.[14] That same year, he went to Wilmington, North Carolina, to work in the Atlantic National Bank under J. Wilkins Norwood, its president, a business and personal friend of his father. His contribution must have been more than satisfactory, for in two years he had been promoted to the office of second vice president.[15]

Will made a fortunate escape from Wilmington to Baltimore in 1897. In the following year, a violent coup d'état took place in Wilmington in which the white population wrested power and property from the blacks who were experiencing considerable economic success and who outnumbered whites. Though Will was not politically inclined, he surely sensed the social tension. More than a year before the violence in Wilmington, the young Coker concluded that banking was not for him. He decided to "follow the bliss" of his childhood and youth, the study of plants. In making this pivotal choice, he gave up a promising future in banking in order to work toward a doctorate in botany at the Johns Hopkins University. His family encouraged him wholeheartedly in this decision.[16]

In a letter of recommendation to President Gilman of Johns Hopkins, President Woodward of South Carolina College called Will "one of the ablest men we ever graduated." His chemistry professor, Dr. W. B. Bur-

William Chambers Coker, the
dissatisfied banker. *Permission
of J. L. Coker II Library, Coker College, Hartsville, South Carolina.*

ney, in recommending him for a scholarship, wrote: "Since graduation,
some three or four years ago, he rose to the second Vice President in one
of the largest national banks in this part of the South. He throws away
these flattering prospects to return to his first love." His history professor
characterized him as "a perfect gentleman . . . an earnest student . . .
with a . . . bent for original investigation."[17] Apparently, while at South
Carolina College Will always tried to read one classic every day in addition to his regular assignments. His particular taste for poetry remained
with him throughout his life.[18]

Arriving at Johns Hopkins at the age of twenty-six, Will became the
first graduate student of Duncan Starr Johnson, the only professor of
botany there. Just five years Will's senior, Johnson had received his Ph.D.
in 1897 and had recently succeeded his own professor, Dr. James Ellis
Humphrey, who died shortly before Will's arrival in Baltimore.

In later years, in response to an inquiry of Dr. Cowles of the department of biology at Johns Hopkins in 1943, Coker had this to say about his
early experience there:

> The first year I got to Johns Hopkins Dr. Brooks sent all of us
> "freshmen" down into the basement room to draw bones of a num-

ber of different mammals, each mammal in a separate box. We were to draw all of the large bones and note the homologies. It was pretty tiresome business and we were down there some several weeks. One of the boys got so bored that he left Hopkins and went to another University. The day after he left Dr. Brooks met me on the stairs and said, "Well, Coker, I see I've got rid of one of you fellows." It was the impression in the department that this tiresome task was put on us to find out who had staying qualities.

When we were drawing fish, which was one of Dr. Brooks' favorite exercises for us, he would often sit by me in the laboratory and discuss the drawings. I would say, for instance, that one didn't look just right to me. I remember Dr. Brooks replied, "Well, it doesn't make any difference whether it is right or not at first, if you can see that it isn't right and correct it."

I remember one of the lab instructors was asked by a student sitting next to me what the creature we were working on was good for. The answer, with considerable emphasis, was, "he is good for himself" which made me realize at once that this instructor was already a good Biologist.[19]

Coker received his Ph.D. in 1901 with a dissertation on seed development in *Taxodium distichum*, the bald cypress, a dominant tree of the swamps and banks of Black Creek near his home in Hartsville, South Carolina.[20] In his dissertation, he thanked his brothers for sending him plant materials for his work. Two of his former students have called his dissertation "a classic study in seed development."[21] The *Botanical Gazette* of the University of Chicago published his paper in 1903 as the first published work from the Johns Hopkins Botanical Laboratory.

William Coker was a loyal alumnus of the Johns Hopkins University. He readily acknowledged the debt he owed to his principal professor, Dr. D. S. Johnson, whose obituary he wrote for *Science* in 1937. For the fiftieth anniversary celebration of the Baltimore university in 1926, he presented a paper and was a contributor to the university's half century fund.[22]

Coker's family and the fellow citizens of his small hometown took note of the remarkable accomplishment of one of its sons. According to family legend, on his return to Hartsville as a doctor of philosophy, he was greeted at the train's arrival by a town band and a chorus of his family and friends singing "Hail the Conquering Hero Comes." Embarrassed, Will descended from the wrong side of the train and made his temporary escape.[23]

Eager to learn more on the general subject of his dissertation, seed development in conifers, Coker sailed for Germany on September 4, 1901 with his sister Jennie, her future sister-in-law Patty Gay, and Dr. James. Will was to work as an independent postdoctoral student in the laboratory of Professor Eduard Strasburger (1844–1912) at Bonn-am-Rhine during the late fall and early spring of 1901–2. Will had referred to the work of Strasburger in his dissertation more frequently than to that of any other writer. He sought to learn more from the most distinguished scholar in his particular field, just as his father had been drawn to the stellar naturalists Asa Gray and Louis Agassiz over forty years before. Strasburger was director of the botanical garden at Bonn at the turn of the century, when Coker arrived to study there. This garden to which Coker had daily access was at that time one of the oldest and finest in Germany. Nearly a century later a list of the interesting plants in the arboretum at Bonn includes an old swamp cypress. Its description is as follows: "Unser Exemplar ist 28.3 m hoch, hat einen Umfang von 356 cm und wurde ca. 1900 geplanzt" [Our specimen is 28.3 m high with a circumference of 356 cm and was planted around 1900]. An old example of the species that was the subject of Coker's dissertation at Johns Hopkins is today prized in the botanical garden at the University of Bonn. It is interesting to speculate whether it was Coker himself who brought the seed or a small tree as a gift to his professor.[24]

In Coker's enrollment book (*Anmeldebuch*) for the winter term, issued at the University at Bonn on October 26, the date of his official entry was November 11, 1901, and the term's end was April 4, 1902. The lectures included three topics of guided independent study with Strasburger, plus self-chosen studies in the history of botany with Professor Noll.*

His sister Jennie spent about three months in Germany travelling from Bonn to nearby sites of natural beauty and literary interest.[25] William worked enthusiastically during his months in Strasburger's laboratory. He said later of his laboratory work with Strasburger, "During the Spring of 1902, while in Bonn, I examined almost daily the maturing pollen grains of a number of conifers, and followed them to the time of shedding."[26]

Coker's postdoctoral studies at Bonn attracted the attention of the

---

* William Coker's enrollment book was issued by the Königlich Preussische Rheinische Friedrich-Wilhelms-Universität zu Bonn, October 26, 1901. He is "Herrn William Coker of Hartsville (America)." This document is preserved in the Coker College archives in Hartsville, S. C. A document signed by the rector of the University at Bonn addresses him as follows: *Iuvenis Praenobilissimus, Guilelmus Coker, Americanus, Studiosus botanicus.* South Caroliniana Library, University of South Carolina, William Chambers Coker papers.

William Chambers Coker, the happy student of botany, possibly as a student at Johns Hopkins or the University of Bonn. *Coker family photograph.*

University of North Carolina's president, Dr. Francis Preston Venable, who himself had studied at Bonn after graduating from the University of Virginia in 1879. Venable later earned degrees in chemistry (M.A. and Ph.D.) from the University of Göttingen, Germany. He respected German-trained scholars. Not having access to funds to attract experienced professors, Venable sought in 1902 to employ promising young scholars.[27] Moreover, Dr. H. V. P. Wilson,[28] professor of biology at the University of North Carolina since 1891, who himself had studied at Johns Hopkins, already knew Coker. Wilson regularly carried on summer studies at the Beaufort, North Carolina marine laboratory.[29] Coker studied the algae of the Beaufort area with his professor at Johns Hopkins, Dr. Duncan Starr Johnson, in the summer of 1899. Wilson, having observed firsthand the quality of Coker's work, doubtless had good reason to recommend him to Venable.

In January of 1902, Wilson wrote Coker in Bonn offering him the position of associate professor in his department. The annual salary was $1,000, and there was no promise of increase in pay or promotion. He was to be responsible for the property of the laboratory and to direct the laboratory work of the class in general biology, meeting eight times weekly. He was to give the lectures on botanical subjects in this class, three lectures weekly for the last two months of the year. Dr. Wilson thought the opportunities for research would be good.[30]

At the beginning of the fall term in 1902, William Chambers Coker began work in the biology department of the University of North Carolina as the first professor trained primarily in botany. He was twenty-nine years old. Being involved in academia in a village surrounded by forests rich in familiar flora must have seemed to him little less than ideal. Chapel Hill was much nearer his cherished home than were Johns Hopkins and the University at Bonn. Both universities, however, had well prepared him for his new adventure as teacher and researcher in botanical studies. Coker was to remain in Chapel Hill in close association with the University for over half a century. Here he would soon make his name as a mycologist.

# CHAPTER TWO

# The Mycologist

*Dans le courant de juin, je partirai pour trois mois, dans le Jura.
Je n'oublierai pas de vous recolter les espèces de nos montagnes.*

*(In June, I leave for three months in the Jura. I shall not forget to
gather for you the species [of mushrooms] of our mountains.)*

—Narcisse Patouillard to W. C. Coker

IN HIS HISTORY of North American mycology, Professor
Donald P. Rogers called William Chambers Coker "the successor, and a worthy successor, of Schweinitz, Curtis and Ravenel."
This ranking places Coker with his own heroes, giants among botanists
of North and South Carolina, from whose lives and writings he constantly gained inspiration. Although best known worldwide for his lifelong work on fungi and water molds, W. C. Coker was by no means ex-
clusively a mycologist. Rogers points to the need to consult the list of
Coker's published research to understand the breadth of his studies,
which include seed development in conifers; native trees, shrubs, and
vines; herbaceous flora; and practical landscape design.[1]

Coker's first mycological publication, with J. D. Pemberton, a paper on
*Achlya*, appeared in the *Botanical Gazette* in 1908.[2] In 1923, the newly
formed University of North Carolina Press published his first two major
works, *The Saprolegniaceae* (water molds) and *The Clavarias of the United
States and Canada* (club and coral mushrooms). His long articles, "The
Amanitas of the Eastern United States" (1917),[3] "Notes on the Lower
Basidiomycetes of North Carolina"(1920),[4] and "Notes on the Thelepho-
raceae of North Carolina" (1921)[5] are classics. His *Gasteromycetes* with

J. N. Couch as coauthor, published in 1928, was widely reviewed and twice reprinted.[6] The *Boletaceae* (1943) and the *Hydnaceae* (1951) were both written with Alma Holland Beers. The latter book appeared just before his eightieth year.[7] Between the years 1908 and 1949, five years after his retirement, Coker was responsible for fifty-nine publications dealing with fungi. Eleven of these are articles of joint authorship. Some are brief notes. Included among his mycological publications are three articles and a bibliography in *North American Flora*[8] and a chapter on the fungi of Venezuela.[9]

Coker was an active member of the national and international fraternity of mycologists of his day, who regularly exchanged information and specimens. His correspondents included mycologists in the United States, Europe, Canada, South America, India, and Japan. These botanists depended upon each other for broader understanding and for completion of works for publication. Coker compiled a long list of botanists that he called his "Exchange List."[10] By continually checking his own observations with other mycologists and by sharing specimens, he sought to increase and complete his registration of specific genera and species. Donald Rogers describes Coker's method as follows: "He kept going simultaneously folders of manuscript on various groups of fungi, publishing papers and books as the material was satisfactorily rounded up."[11]

### Coker's North American Mycological Correspondents

On the faculty at Harvard, William Gilson Farlow (1844–1919) was the first professor in America of cryptogamic botany, the study of spore-bearing plants, such as ferns, mosses, algae, and fungi. He remained at Cambridge from the time of his appointment in 1879 until his retirement in 1896. At the time of Farlow's death in 1919, the cryptogamic herbarium at Harvard consisted of several hundred thousand specimens. At present, the Farlow herbarium contains almost 1.4 million specimens.[12] It is interesting to note that Farlow was from 1870 to 1872 assistant to Asa Gray (1810–88), the professor of natural history at Harvard who had inspired Will's father, James Lide Coker, during his studies there in the 1850s.

There are preserved twenty letters between W. C. Coker and Farlow, dated from January 28, 1916, to March 1, 1919, three months before Farlow's death.[13] They reveal a friendly professional relationship. During this period, Farlow sent Coker specimens of *Amanita*. At Farlow's request, Coker sent him seed of sheep laurel (*Kalmia angustifolia*) through the kind offices of his cousin Paul Rogers in Hartsville. Farlow invited Coker to work in the Curtis herbarium. Coker hoped that Dr. Farlow would visit him in Chapel Hill, but the elder botanist was unable to accept his invitation.

Between March 5, 1920, and November 23, 1920, after Dr. Farlow's death in June of 1919, Coker corresponded with Dr. Lincoln Riddle and Dr. Roland Thaxter, both also of Harvard's herbarium of cryptogamic botany, about specimens in the Curtis herbarium and about the index of North American Fungi. Apparently, he visited the herbarium at Harvard more than once during 1920.

One of Coker's most faithful and helpful correspondents on mushrooms was the North Carolinian Henry Curtis Beardslee (1865–1948), instructor at the Asheville School for Boys. Fifty-one letters passed between Beardslee and Coker from February 1915 through February 1920.[14] Working together, the two mycologists sought to complete the identification of North Carolina mushrooms. They exchanged specimens regularly and generously, with mutual enthusiasm. On June 1, 1918, Coker reported to Beardslee in excited prose that he had received a photograph from Mrs. Jervey of Charleston of an *Amanita* that "I am quite sure is your new species." He promised him a dried specimen, if he could possibly obtain one.

Genera of fleshy fungi most frequently mentioned in the correspondence between Coker and Beardslee are *Collybia, Russula, Clavaria, Amanita, Boletus, Tremella, Dacrymyces, Exidia,* and *Hydnum.* Beardslee was co-author with Coker of three articles: on *Collybia* in 1921, on *Lactarias* and *Clitocybe* in 1922, and on *Mycenas* in 1924.[15] In January 1920, Coker mentioned to Beardslee the *Exidia* that grows on *Robinia* (locust), a fungus that twenty-four years later would be named in Coker's honor.[16]

Between February 6, 1915, and January 20, 1920, there was an exchange of eleven letters between William Chambers Coker and Calvin H. Kauffman (1869–1931), professor of botany and curator of the cryptogamic herbarium at the University of Michigan, Ann Arbor. They chiefly discussed species of *Cortinarius* and *Inocybe.*[17]

Between February 1916 and November 1920, Coker and Edward A. Burt (1859–1939), director of the Missouri Botanical Garden in St. Louis, exchanged twenty letters and many specimens. This correspondence dealt with a dozen or more genera of fungi.[18] In a letter of December 1919, Burt identified a long list of Coker's specimens. There is some discussion in Coker's reply of December 22, 1919, about Burt's determinations.[19]

George F. Atkinson (1854–1918), professor of botany at Cornell, was familiar with the Carolinas. He had taught in the 1880s at both South Carolina College, Coker's alma mater, and the University of North Carolina.[20] Apparently Atkinson played an important role in awakening Coker's interest in mushrooms. John N. Couch and Velma Matthews wrote that Coker arrived at Chapel Hill with a copy of Atkinson's recently published *Studies of American Fungi* under his arm.[21] Almost immediately Coker set out to collect mushrooms that sprang up in damp spells on the forest floors and meadows around Chapel Hill. It is easy to see how Atkinson's beautiful book, with its full-sized photographs, some in color, could send this young botanist, eager to know the fungi of his new surroundings, into the woods and fields to collect them for himself.

Less than a decade after Coker's early collections at Chapel Hill, there began a lengthy correspondence between Coker and Atkinson. A letter from Atkinson, dated November 28, 1910, informed Coker that he would be glad to look over his fungi.[22] Between February 1915 and August 1918, twenty-four letters chiefly concerned with species of *Clavaria, Amanita,* and *Collybia* passed between Atkinson and Coker. On March 9, 1915, Atkinson offered to take Coker with him on mushroom field trips around Ithaca during the coming summer.[23]

Atkinson's letters to Coker were among his last. Before starting his cross-continental trek to collect mushrooms in Mt. Rainier National Forest, Atkinson wrote Coker asking for advice about stopping points as he passed through the South before leaving for the state of Washington.[24] In another letter he expressed his desire to visit Chapel Hill during some long rainy period.[25] Apparently he did visit Chapel Hill on his southern trip and collected with Coker there, but because of heat and drought, their collection was disappointing. He returned to Ithaca to prepare to leave earlier than anticipated for the west coast.[26] While collecting near Mount Rainier later in the fall, the distinguished mycologist became seriously ill. He died of pneumonia in mid-November of 1918, probably a victim of the deadly worldwide strain of influenza of that year. Atkinson literally lived and died with his fungi. On November 19, 1918, the American cryptogamist Theodore C. Frye (1896–1962) wrote Coker in great detail about Atkinson's mortal illness contracted during the collecting expedition in Mount Rainier Park, a foray that he had anticipated with such enthusiasm. "Shortly before he died," Frye wrote, "he was trying to dictate to his nurse some notes on his fungi. . . . Thus he died practically working on his fungi even though he was delirious."[27] Later, Coker received several letters from Atkinson's associates at Cornell, chief among them Dr. H. H. Whetzel of the botany department, about the future of Atkinson's unfinished ten-volume work on mushrooms.[28]

William Alphonso Murrill (1869–1957) was the assistant director of the New York Botanical Garden and editor of the journal *Mycologia* for sixteen years. There was an exchange of twenty-three letters between Coker and Murrill from February 18, 1915, to April 21, 1920.[29] They sent each other specimens of fungi for identification and shared information on the genera *Typhyla, Clavaria, Amanita, Mycena, Omphalia, Tricholoma, Amanitopsis,* and *Tremella.* On September 24, 1915, Coker excitedly reported to Murrill an episode of unusually successful collecting. "On returning home," he wrote, "I found a full crop of mushrooms coming on and have been busy trying to keep up with them. A good many additions to my list turned up."

Perhaps sensing trouble for himself at the New York Botanical Garden, Dr. Murrill wrote on February 13, 1919, asking if Coker could use him as a research professor at Chapel Hill. He told Coker in confidence that he was giving Coker the first chance at himself. The mutual trust of the two men was clear in this letter, which is perhaps a harbinger of more serious difficulties ahead for Murrill, who in 1924 as a very sick man permanently left the New York Botanical Garden.[30] Later, Coker corresponded with Murrill after he had become a professor of botany at the University of Florida in Gainesville. The length and depth of Coker's friendship with Murrill is obvious in Coker's greeting, dated October 10, 1944, on the occasion of Murrill's seventy-fifth birthday: "Hail to you tall friend, gracious, honest, and always generous, indefatigable student and lover of nature. When I recall in the sunny lanes of memory your kind and cheerful face, I have my hope renewed that this sad world is not so wicked as it seems to be. May you live a long and happy life, dear friend."[31]

There is preserved an interesting correspondence between Coker and Curtis G. Lloyd (1859–1926), director of the Lloyd Library and Museum in Cincinnati, Ohio. The Southern Historical Collection contains eighteen letters between Coker and Lloyd from July 1917 to August 1920. Apparently Lloyd was well known as an authority from whom mycologists sought information on mushrooms and without whose advice they dared not publish a species that the petitioner considered a new discovery. Lloyd, however, in his correspondence with Coker, reveals himself as crusty, not to say curmudgeonly. The two men always couched their most critical comments between an introductory pleasantry and a closing of gentility, the usual epistolary etiquette of the traditional "scholar and gentleman" of the day. Lloyd, however, often permitted himself impertinent and even insulting comments in the central part of his letters. In a letter dealing with the name of a species of *Dacrymyces,* Lloyd wrote to Coker on July 17, 1920, "I am sorry to see you such a close follower of heresies of Kuntseism for in some cases it has led you into taking grounds that are going to be embarrassing to you." He then proceeded sarcastically to accuse Coker of representing that Professor Peck was the first to name the species in question. He added, "I imagine the mycologists of France who have collected and recorded this plant perhaps hundreds of times in their fungus forays would be interested to know that they must come to America to get names for the common plants of France."[32] Lloyd followed protocol by closing his letter on a courteous but boastful note, inviting Coker to go through his collection in Cincinnati.

Two drafts of Coker's response to this acerbic attack are preserved.[33] In the first version, dated July 23, 1920, which Coker did not mail, he began with a graceful comment, veiling a rebuke: "As you are usually admitted to be a privileged character, I am not impatient with your criticisms which, in my opinion are more foolish than wise." He added, "You very carelessly mistake the facts, etc." He then proceeded vigorously to defend himself from unjust accusations. In a cooler version written July 31, a week later, Coker modified his language somewhat while still keeping his spirited defense against Lloyd's unfair and inaccurate criticism. Coker closed his second letter with the usual compliment, thanking Lloyd for his invitation to look over his collections and saying that this would give him pleasure and adding that he hoped some day to avail himself of the offer. In all of Coker's correspondence that I have read, this letter in response to the crudely inaccurate accusations of the crusty Lloyd comes nearest to a departure from Coker's usual courteous style. Though tempted to answer in kind, Coker nonetheless respected the limits of gentility he set for himself.

Respecting collectors everywhere, Coker was eager to learn from any source that came to his attention. His sensitive treatment of amateur aspiring mycologists was in stark contrast to the high-handed approach of Lloyd, whose lack of regard for the feelings of others is never more evident than in some apparently disparaging remarks to Miss Ann Hibbard of Roxbury, Massachusetts. Lloyd's comments to Hibbard seem to have caused her to doubt her own identification of collected specimens. Miss Hibbard was an enthusiastic amateur collector and illustrator of *Clavaria*, a genus that was the subject of one of Coker's major books of 1923. Between September 1919 and April 1920, there was an exchange of nine letters between Coker and Hibbard.[34]

On the advice of Murrill, Coker wrote Hibbard, informing her of his proposed publication on *Clavaria* and asking to see her specimens and colored drawings. He requested permission to keep one specimen where there were duplicates and return the rest to her. Miss Hibbard sent him a box of specimens of *Clavaria* for him to identify. She included five watercolor drawings. She asked for some specimens of *Clavaria* in return, listing the species she wanted. Without being condescending, Coker told her in his reply that five of the species she had sent were those she had requested of him. He then respectfully asked permission to use some of her colored drawings in his coming publication, of course giving her full credit. In reply she requested that he not use two of the drawings, since she was uncertain of the species. Coker encouraged her by asking why he could not use her drawings of the plants, if he mentioned her opinion, and why she considered these two specimens different from the others. She responded by describing the differences of her specimens from *Clavaria cristata* and added, "I am writing the above with some reluctance because I feel that I am not qualified to back up my opinions. They are only my 'guesses' as Mr. Lloyd puts it." Two more of Coker's letters to Hibbard either identify her specimens or express his own uncertainty. He considered her observations as valid as his own.[35]

Coker learned much about *Lactaria* and *Russula* from Gertrude Simmons Burlingham (1872–1962), Ph.D. from Columbia University (1908), a longtime teacher of botany in a Brooklyn high school. Thirteen letters passed between them during the period of October 1915 to May 1918, when Coker was preparing to publish an article on the *Lactarias* of North Carolina.[36] As usual, he sought help from a known authority, in this case Burlingham, to check the accuracy of his work.

Mrs. Ida M. Jervey of Charleston, S.C., an amateur collector, corresponded with Coker about *Amanita*. At least six letters passed between them from early 1916 to mid-1917. She sent him descriptions and drawings, and he sent her his *Amanita* paper, which was published in 1917 in the *Journal of the Elisha Mitchell Scientific Society*. Once, he felt the need to caution her about tasting specimens of *Amanita* to test their edibility (some species are deadly).[37]

In three letters to the director and curator of the Marine Biological Laboratory at Woods Hole, Massachusetts, during June and July of 1920, Coker proposed an exchange of specimens of the water molds *Saprolegnia* and *Achlya*. In March of 1917, he received permission from Howard Banker of the Eugenics Record Office at Cold Spring Harbor to publish the first species to be named for him, *Phellodon cokeri*.

### Coker's Foreign Mycological Correspondents

Coker did not restrict himself to American colleagues in his mycological correspondence. He eagerly exchanged information with European mycologists after World War I (1914–18). On January 24, 1920, he wrote to Professor Louis Mangin (1852–1937) of the cryptogamic laboratory of the Museum of Natural History in Paris about the species *Dacrymyces deli-*

*quescens* Bulliard.[38] In his reply (in French) on February 23, Mangin sent greetings to H. R. Totten, Coker's student who, having elected to spend some months studying in France after his military service in 1917–18, had called on Mangin the previous spring.

From February to October of 1920 several letters passed between Coker and Professor Léon Dufour (1862–1942) of the science faculty at the Sorbonne and the laboratory of vegetal biology at Fontainebleau.[39] Dufour thanked Coker for sending him his articles. He spoke of writing résumés of Coker's and Beardslee's work on *Amanita, Lactarius, Russula, Hydnum* and other genera for the *Revue Générale de Botanique*.

The French botanist remarked that the appalling experiences they had just endured had at least the good consequence of their own communication and that they should continue to work for the advancement of science and for civilization in general. He offered to send Coker his own modest book on useful knowledge of the mushrooms and one on the laboratory of Gaston Bonnier (1851–1922), where he was assistant director.[40] He remarked on H. R. Totten's happy memory of his visit to his laboratory and assured Coker that other Americans would receive a cordial welcome there. Coker replied by thanking him for the gift of his book on mushrooms and congratulating him on its appearance and usefulness. On a political note, Coker also expressed his own conviction that "the Senate's rejection of the Peace Treaty is condemned by the great majority of American citizens." He commented, "Petty politics and partisan motives were at the bottom of this betrayal of the interests of the people." Coker said he hoped to meet Dufour on his next trip to Paris. On September 12, Dufour wrote him that he had not been able to find the mushroom he desired but would call it to the attention of the members of the society. On October 27, 1920, Coker asked Dufour the favor of securing for him one or more good specimens of *Naematelia encephala* Fr. He offered to send Dufour any available specimens of fungi he wished.

On November 6, 1920, Dr. Christian Glück (1868–1940), professor of botany at Heidelberg and a specialist in aquatic plants, though not a mycologist, wrote asking that Coker send him material including "seeds, knobs or living shoots" of about seventeen species, while offering him in return dried herbaceous plants of the continent of Europe. On May 28, 1927, Hans Sydow (1879–1946) of the *Annales Mycologici* in Berlin wrote Coker that he was eager to exchange specimens to enrich his herbarium of fungi. He told Coker of possessing his "great works," the *Saprolegniaceae* and the *Clavarias*, and that he would accept any other reprints of his works in exchange for fungi.

Between January and November of 1920, at least five letters passed between Coker and the famous French mycologist-pharmacist Narcisse Patouillard (1854–1926) of Neuilly near Paris. On January 20, 1920, Coker wrote Patouillard mentioning three fruitless attempts of his student H. R. Totten to meet him during May and June of 1919. Coker asked for specimens from the Patouillard herbarium, especially species of *Clavaria*, as well as species of *Tremella* and *Dacrymyces*. He told Patouillard of his having written to Dr. Mangin about *Dacrymyces deliquescens* and inquired as to whether in his opinion this fungus grew on coniferous trees. On April 2, 1920, Coker thanked Patouillard for sending him his specimen of *Dacrymyces deliquescens* and declared the difference between himself and Europeans to be that of description of habitat. Coker thanked Patouillard for his offer to collect and send him specimens of French *Clavaria* and *Tremella*, among other fungi. He included the same apology for the U. S. Senate's action on the peace treaty, which he had earlier sent to Dufour. Patouillard replied on May 29 thanking Coker for sending him specimens and descriptions. He declared *Platygloea caroliniana* to be a separate species from *Heliogloea lagerherini*, as he found *Heliogloea* to be a bad genus. He said he was leaving for three months in the Jura where he would be glad to collect mountain specimens for Coker. On October 26, 1920, Coker wrote to Patouillard telling him that he was sending him the most recent number of the *Mitchell Journal*. He expected to publish another chapter on the fungi of North Carolina in the next number. He asked for a good specimen or two of *Naematelia encephala* Fr. Patouillard replied, describing *Tremella encephala* and offering to share with Coker a piece from his herbarium. He included in his letter sketches from a previous study of this species. On June 9, 1927, Dr. E. Gilbert in Paris wrote to thank Coker for the specimens of *Amanita*. Marcel Josserand of Lyon, France, sent Coker pictures of his specimens of *Clavaria*.[41]

Coker initiated a correspondence with Lars Romell (1854–1927) of Stockholm, Sweden, on February 24, 1920. He explained his work on the *Clavarias* for *North American Flora* and also the large volume with many plates that he expected to publish. He asked for European specimens of *Clavaria*, mentioning twenty-four species, four of which he considered particularly important. He offered to pay for the favor and also to furnish Romell with specimens that he could supply. Coker mentioned that the North Carolina botanist Beardslee had visited Romell in Sweden and that it was Beardslee who had suggested that he (Coker) write to him about these fungi. In a reply dated April 16, 1920, Romell attributed the delay in his response to the situation in Germany. He described certain *Clavaria* that he was sending but claimed to be no expert and said that some of the information could be unreliable. He requested that some specimens be returned. On May 11, 1920, Coker replied to Romell's letter of April 16, 1920, telling him that he was awaiting Romell's materials and that he was sending him a small box of resupinate *Hydnaceae* and *Polyporaceae* as requested.[42]

Later Coker wrote Romell that the expected box of most valuable specimens, drawings, and notes had finally arrived and had been of great help to him. He would return those parts that Romell wished to retain. He remarked that he was sending him the last three numbers of the *Mitchell Journal*. He offered to send any specimens available that Romell cared to have. Coker himself requested one or more good specimens of *Naematelia encephala* Fr., a fungus that grows on pine.[43]

Coker and the Swedish mycologist became good friends. This friendship was cemented when Coker met Romell and worked with him in the Bresadola herbarium in Stockholm in the summer of 1921. Coker wrote to Romell about *Amanita* on July 18, 1927, less than a month before Romell's unexpected death. At the request of Romell's son, L.-G. Romell, Coker wrote his father's obituary for the *Mitchell Journal*.[44] In a letter dated October 29, 1927, he inquired of the younger Romell about the dispensation of his father's library and herbarium, the one "in his office."[45] Coker apparently did succeed in purchasing for the botany library at Chapel Hill some of Romell's most valuable books. Mycological books with Romell's autograph on the flyleaf may be studied today in the rare book room in UNC's Wilson Library and in the library's general collection.

Coker was in correspondence with Dr. John Dearness (1852–1954) of London, Ontario, in the effort to secure clavarias from his region. In a letter of March 24, 1920, Dearness offered to send Coker what he might want from the catalog of the species collected in his locality. Coker replied, asking Dearness about *Clavaria spinulosa* and requesting a collection of this species. Dearness confessed to having lost his letter but mentioned *C. spinulosa*, adding that he had sent his specimens to Dr. Robert A. Harper (1862–1946), who could suggest where they could be found.

On March 30, 1920, Coker wrote to Mr. Carleton Rea (1861–1946), secretary of the British Mycological Society, to ask for an exchange of its *Transactions* with UNC's *Journal of the Elisha Mitchell Scientific Society*, a publication in large part concerned with mycology.

William Chambers Coker corresponded with other distinguished mycologists in Europe, Japan, and South America, among them the Italian botanist Abbé Bresadola (1847–1929) of Trento, Italy, to whom he sent a box of plants in 1922 and whose herbarium in Stockholm he had visited the previous summer.[46] Others of his correspondents include Masaji Nagai (1905–66) of Sapporo, Japan, and Carlos E. Chardon (1897–1965) of Caracas, Venezuela.*

Coker spared no effort to broaden his knowledge of fungi from all possible sources. A faithful correspondent, he sometimes suffered frustrating delays and lost letters, especially in the years following the end of World War I in 1918, when he was working to complete his two major books on fungi that UNC Press would publish in 1923. Sometimes, as noted above, letters from mycologists informed him that a specimen had been lost or that the political situation had caused delays. Nonetheless, he persisted in seeking the information he needed. Cordial rapport, frequent correspondence, and exchange of specimens among mycologists at home and abroad allowed him to check and double-check his own observations on fungi and thus to publish full and accurate information.

Coker achieved scholarly eminence as a mycologist; yet mycology was far from being his only botanical interest. As we shall see in the next chapter, he loved being in the field, and everything that grew there attracted his eye and mind.

---

* Other foreign correspondents of Coker's during the years between 1927 and 1937 are Sultan Ahmad of Ladhar, Sheikhupura, Punjab (India); Miss Catherine Cool of the Rijks Herbarium in Leiden, who wrote about the identification of *Clavaria*; Dr. Charles Cejp of the Botanical Institute at the Charles University in Prague; Professor Sanshi Imai of the agricultural faculty of the Botanical Institute in Sapporo, Japan; Dr. Ivan Klástersky, director of the botany section of the Národní Museum in Prague, who requested the exchange of "doublets," especially *Polyporaceae* and *Thelephoraceae*; the Rev. Father J. Rick, Seminario Sao Leopoldo, Rio Grande do Sul, Brazil; Dr. Vandendries of La Chanterelle, Rixenart near Brussels; and Dr. Joha Westerdijk, director of the Phytopathological Laboratory "Willie Commelin Scholten" in Baarn, The Netherlands. SHC.

# CHAPTER THREE

# The Field Botanist

*Field trips I took with him were always a wonderful
thing, because he had . . . an instinct for discovering
where things were, whether it be a mushroom, a
flowering plant, a bird. . . .*

—Paul Wilson Titman

THE BOTANICAL INTERESTS of William Chambers Coker became all-consuming from his childhood forward and, to judge from his publications, eventually nearly all-encompassing. From his early years as a professor in Chapel Hill, North Carolina, Coker collected mushrooms and other fungi. He also found time to study trees, shrubs, vines, and herbaceous plants. His first tree book with H. R. Totten, *The Trees of North Carolina*, appeared in 1916. Continuing to study trees, he and Totten published in 1934 a much-expanded edition, *Trees of the Southeastern States*, and yet another edition in 1945.

During his summers in Highlands, North Carolina, W. C. Coker led frequent botanical expeditions. His companion on one of these trips was Henry Wright, a former U.S. Forest Service employee who had a vast knowledge of mountain plants. Coker and Wright sought to find the site where Asa Gray had rediscovered *Shortia*, the Oconee bell, first collected by André Michaux and subsequently lost to botanists for many years.* On

---

* *Shortia* is now commonly recognized in the wild and is available in the plant trade. Radford, Ahles, and Bell's *Manual of Vascular Flora of the Carolinas*, 2nd ed. (1968) 818. The *Manual* reports *Shortia* as growing in three North Carolina counties and three South Carolina counties. For the story of the lost *Shortia* and Gray's search, see Charles F. Jenkins, "Asa Gray and his quest for Shortia galacifolia," *Arnoldia*, 51.4 (1991) 5–11.

William Chambers Coker
before a specimen of *Arbor
vitae* on a field trip on May
14, 1926. *Southern Historical
Collection, Wilson Library,
University of North Carolina
at Chapel Hill.*

another occasion Coker led a field trip to the olivine-serpentine barrens in
Clay County, North Carolina, where he encountered plants seldom found
in the state.[1] He played a major role in the 1933 Highlands summer foray of
members of the Mycological Society of America.[2]

One of Coker's students, Paul W. Titman, who later became a professor of biology in Chicago, said of his former teacher, "Field trips I took
with him were always a wonderful thing, because he had, one might almost say, an instinct for discovering where things were, whether it be a
mushroom, a flowering plant, a bird, whatever."[3]

Titman recounted an apocryphal tale circulating among botany students, one which Coker himself would have found amusing: "Once Dr.
Coker took a small nephew on a field trip, stopped, and fed him pieces of
a mushroom. After an hour, he asked the child, 'How are you feeling?'
'Fine, Uncle Will.' Whereupon Dr. Coker exclaimed enthusiastically, 'Splendid, another edible mushroom.'"[4]

Another of his students, Albert E. Radford, author along with Harry E.
Ahles and C. Ritchie Bell of the *Manual of the Vascular Flora of the Carolinas*, remembered that before accepting him as a graduate student in

botany, Coker drove to Furman University in Greenville, South Carolina, where Radford was a student. There Coker tested the young man's interest and powers of observation on a field trip, presumably to assure himself of his capability and to emphasize for the prospective student the serious nature of the work he would undertake.[5]

### Early Ecologist

Coker's interest in ecology, evident to his field trip companions, began early in life. The first of Coker's books, *The Plant Life of Hartsville*, published in 1911 with an enlarged edition in 1912, is based on his youthful study of plants around his home.[6] As an introduction to his discussion of the plants of Hartsville, Coker recorded the climate, altitude, latitude, distance from the sea, temperature, and annual rainfall of Hartsville as compared to those of nearby towns. He described each of the six ecological areas of distinct plant communities around Hartsville: the sandhills, the well-drained upland forest, the poorly drained flatwoods, the savannas, the swamps, and the streams and ponds.[7] He indicated which plants were native to each area, when they bloomed, and when they were in fruit.

Wherever the passionate botanist happened to be, he closely observed and carefully recorded the plants around him. He collected and mounted dried specimens. Each specimen sheet recorded the location and date of collection and thereby documented his travels and his visits to family members. He daily registered noteworthy characteristics of plants he encountered on his routine campus strolls. He reported one of these observations in an article on oak seedlings: "There is in Chapel Hill, N.C., a magnificent tree of *Quercus alba* L. that shows the same [multiseeded] peculiarity. Through a number of years I have watched this tree and there are always a large proportion of its acorns that contain two or three young plants."[8] Little escaped his eagle eye. In September of 1917, Coker noticed a bed of mandan corn near the entrance of one of the buildings of the New York Museum of Natural History. He wrote asking the secretary of the museum to refer him to any relevant publications. On being informed that the Montana Agricultural Experiment Station had published a bulletin on this species, he immediately ordered a copy and on the same day wrote to thank his informant at the museum for the valuable information.[9] Handwritten notes scrawled on field trips attest to his observation of the local flora, whether he was climbing Black Mountain or tramping in Kanuga forest. During visits to Hartsville, South Carolina, and Highlands, North Carolina, he observed all manner of plants. By experiments to ascertain the limits of conditions in which plants would survive, if not flourish,

Nancy Eliason and S. O. Trenthon, beside the botany touring cars "Amanita" and "Boletus" on a field trip in eastern North Carolina, May of 1926. *Southern Historical Collection, Wilson Library, University of North Carolina at Chapel Hill.*

Coker tested nonnative plants in his own garden and in the Arboretum on the UNC campus. Finding some of these plants suitable for the local climate, he promoted their use in the locality.[10]

The epigraph to his first book, *The Plant Life of Hartsville,* is a quotation from the eighteenth-century English clergyman and naturalist Gilbert White: "It is I find in zoology as it is in botany; all nature is so full that that district produces the greatest variety which is the most examined."[11] Coker probably chose this sentence to justify the botanist's intense scrutiny of any small area. His choice also reveals his own love for the flora of his home and his awareness of the inexhaustible offerings of nature to be observed there.

### Determining the Range and Stations for Native Plants

Coker had an enduring interest in correcting mistakes about the range of native plants and in solving problems of their location. Citizens of his home of Darlington County, South Carolina, and indeed botanists in general, gave the common name "Darlington oak" to *Quercus laurifolia* on the assumption that the tree was native to the town. To the consternation of some of his friends and kinfolk who took pride in the name, Coker, finding no old laurel oak trees, judged that someone had planted several specimens at some time in the past and that the oak had then established itself in the wild.*

Upon hearing in the spring of 1919 of a *Rhododendron* blooming at Selma, North Carolina, Coker immediately set out to verify the report for himself. He wrote, "Driving out to the banks of the Neuse River about 4 miles north of town [Selma], I found *Rhododendron catawbiense* in full bloom. This record extends the species to the middle of the coastal plain and down to an elevation of about 150 feet above the sea."[12]

William Coker's intense interest in the location of the plants of the Carolinas prompted him to send to the *Times* of Georgetown, South Carolina, on November 5, 1920, the following notice: "Wanted: Information about Venus' Flytrap.[13] One of the most remarkable plants in the world [is] called Venus' Flytrap because it catches and digests living insects. Was reported from near Georgetown many years ago by Elliott in his 'Sketch of the Botany of South Carolina and Georgia.' No specimen from South Carolina is now known. Information in regard to the present occurrence of this plant in South Carolina is greatly desired. Address W. C. Coker, Professor of Botany, University of North Carolina, Chapel Hill, N. C."[14]

On reading a statement of Dr. Bashford Dean, published in 1910, that a Mr. Walter Hoxie of Beaufort, South Carolina, had seen the venus flytrap as far south as St. Augustine, Florida, Coker decided eighteen years later to investigate in person this incredible claim. He told the story in his own dry style: "After some correspondence I went to Savannah to see him (Hoxie) and found that if he had ever seen *Dionaea*, which was very doubtful, he could not now find it."[15] Twenty years later, in 1948, C. R. Bell found venus flytrap south of Georgetown, South Carolina.†

Between May 1932 and June 1935, Raymond F. Ashley of Ashe County, North Carolina, sent Coker several specimens of an extraordinary *Rhododendron*. In June of 1935, Coker visited Ashley and photographed the plant. He made a careful study of one plant in flower and upon finding it to be unique, honored the collector as discoverer. He wrote, "We take pleasure

---

* See Coker's article, *JEMSS* 32.1 (April 16): 38–40. W. C. Coker's brother, James, planted two rows of these oaks from the acorns of one tree just west of the old Coker home place. These trees, now of considerable size, border Laurel Oak Street in Hartsville, South Carolina. (This information is derived from the author's conversation with David R. Coker, who suggested to his daughter a study of the variations in these trees, derived from a single parent.)

† On June 20, 1948, twenty years after Coker's trip to Savannah, C. R. Bell found a station of venus flytrap in Charleston County, about fifteen miles south of Georgetown. See hand-drawn map by Bell in the Coker College archives. As Bell had been a student of Coker's, his professor doubtless knew of this discovery.

in naming this plant for Mr. Raymond F. Ashley (*Rhododendron Ashleyi*), who first sent us a twig and leaves in May, 1932; another in September, 1932; fading flowers in July, 1934 and fresh flowers in full bloom in June, 1935."[16]

## Coker's Field Research Techniques

Coker tackled the problem of whether the *Magnolia cordata* or *Tulipastrum cordatum*, as named by his revered hero André Michaux,[17] could be considered a separate species. His article, published in 1943, demonstrates the thorough manner in which Coker conducted his own research. He reviewed the literature, quoting twenty-two publications that treat this small, native, yellow-flowered tree. He recorded his own personal observation in the field in April of 1942 when, in the company of Professor Edward Caleb Coker of the University of South Carolina[18] and other botanical enthusiasts, he sought to verify examples of this plant reported to be in the vicinity of Columbia, South Carolina. He arrived at the "disappointing conclusion that *cordata* is merely a rather vaguely defined, marginal extension of the southern yellow-flowered variety of *M. acuminata*." He added, "We are evidently dealing with a very variable group with extremes connected by a series of forms. This being so, we can hardly be justified in calling the lower piedmont (fall line) tree anything more distinct than a variety."[19] He then described the plant carefully to buttress his own opinion. The article concludes with a bibliography and a map showing authenticated entries made in the field or records of plants then in herbaria. W. C. Coker was thus ready to debunk, to verify, to record, or to modify the reported range and identity of unusual plants. He went to considerable physical and intellectual effort to note the extent of floral habitat and to correct erroneous information.

In his tenacious effort to establish the range of native plants, Coker did not hesitate to send associates and friends on difficult trips in pursuit of a plant he was studying. While away from Chapel Hill, and eager to complete an article in preparation,[20] he wrote his assistant, Miss Alma Holland, three urgent letters instructing her to verify the incredible report that a *Rhododendron* native to the mountains was growing in sandhill country near the border of North Carolina and South Carolina. On July 10, 1919, he issued Miss Alma his first order on the subject: "Get Mrs. Russell to send specimen of Rhododendron from Richmond county as soon as possible." He stepped up the urgency on July 26. "Do not let the Rhododendron matter rest . . . As soon as the leaves arrive from Mrs. Pegues, you will see whether they are Rhododendron or not and let the map go on

Alma Holland Beers beside blooming wisteria. *Courtesy of the Couch Biological Library, Coker Hall, University of North Carolina at Chapel Hill.*

[to the publisher]." Apparently not hearing of the arrival of the specimen, Coker instructed Miss Alma to visit the reported site in person and "get a good collection." He instructed her to take the train south from Raleigh or Apex and added, "The train leaves Raleigh at 5 AM." [21]

Eager to verify a report that the venus flytrap grew as far south as the "big ocean" bay near Georgetown, South Carolina, he asked his friend, the poet Archibald Rutledge, who lived in the area, to find it at this site for him. Rutledge reported on May 1, 1938, that he had not been able to locate the plant, though he himself knew it was there. He volunteered to try again, while adding the offhand comment that in his search he had encountered a diamondback rattlesnake. Even mention of the rattler did not get Rutledge off the hook. Coker replied to his letter on May 5 that he hoped his friend did not get bitten and that he would appreciate a line on that subject. He continued, however, by reminding Rutledge of the urgent question: "You may be sure I am looking forward with much interest to getting the venus' flytrap from you, and I greatly appreciate your continued interest." [22]

Another example of Coker's assigning to friends collecting tasks in less than comfortable circumstances was his request of Frank Tarbox, horticulturalist for Brookgreen Gardens, to go to the eastern bank of the Waccamaw River "at low tide" to try to find the Ogeechee lime (*Nyssa ogeche* Marshall; *N. capitata* Walter), reported to be one mile or less up the river from Brookgreen. The muddy banks of the Waccamaw are home not only to a rich variety of plants but also to interesting animal life most active in the hot summer of the area. Here water moccasins festoon branches over the black river, alligators sun themselves on muddy banks, and the drone of biting insects persists. A botanical adventure at certain times of the year was not without its hazards. Coker's letter, written April 22, 1943, arrived early enough in the year for Tarbox to avoid encounters with these denizens of low country swamps. He was obliged to plan a second effort, however. Two weeks later Tarbox wrote Coker, "I took a trip out on the river this afternoon, and I looked closely for the Ogeechee lime, but did not see anything which I thought looked like it. Perhaps I can locate it, if it is there, better when the fruit is about grown."[23] He added that he was writing Mr. Jones, Coker's source for the information, for a more definite location so that he might find the plant.[24] On May 24, 1943, Tarbox reported that he had heard from Mr. Jones, who reminded him that Tarbox himself had been along on the field trip when they had spotted the plant. Though Tarbox remembered tramping the woods, he had no recollection of the tree, but he assured Coker that during Jones's visit to Pawleys Island in June, Jones would take him to the area to show him the tree.

Coker became adept at delegating, one might even say commandeering, friends who shared his interests to collect specimens for him, especially during the years 1943 and 1944, a time of stringent wartime gasoline rationing. Frank Tarbox was a faithful collector for his friend in their joint effort to gather all possible material for Coker's projected article on native smilax vines.[25] Between September 1937 and July 1943, twenty-three letters between Coker and Tarbox dealt with the collection of various species of smilax in the region of Brookgreen Gardens. Coker needed specimens of leaves, stems, roots, and male and female flowers of each variety.

Another of Coker's smilax-hunting allies was G. Robert Lunz, one of three acting directors of the Charleston Museum during the absence of the director, E. Milby Burton, who was called to war service. Though a marine biologist, Lunz was a friend of Coker's and a willing collector of plants. Between January and May of 1943, ten letters on the collection of varieties of smilax passed between the two men. Coker instructed Lunz to dig and mail to him roots, leaves, and stems of *Smilax glauca* and to collect flowers, press them, and send them to Chapel Hill. He also asked Lunz questions on the local use of smilax tips for food and on their more general use in making "Jack briar" pipes.[26]

Determined as always to find the most solid reference, Coker requested his friend and former student H. A. Allard to check the four smilax plates in the first edition of Mark Catesby's *Natural History of Carolina, Florida, and the Bahama Islands*, 1731, 1743, plates 15, 47, and 52 in volume I and plate 84 in volume II. He wanted to be sure that the descriptions on the plates in the 1771 edition in the University library at Chapel Hill did not differ from those of the first edition. Calling all local institutions in the Washington area, Allard discovered that the Library of Congress possessed a Catesby first edition. He was disappointed to learn, however, that the two precious volumes were then in protective storage at the University of Nebraska for the duration of the war. However, Allard was able to check the smilax plates in the 1754 edition of Catesby in the Congressional Library's rare book room. He copied and sent to Coker Catesby's botanical names for all four species in this second edition available to him. Coker thanked Allard for consulting a source nearer to the original than that available at Chapel Hill.

As it turned out, plate numbers and botanical descriptions of the four varieties of smilax depicted in Catesby's *Natural History* are identical in the 1754 edition and in the first edition.[27] While taking a risk, Coker did not err in listing the 1731 and 1743 volumes in the bibliography for his 1944 smilax article. Though he was well aware by consulting subsequent editions that Catesby had represented three birds and a moth with four different species of smilax and though he knew their plate numbers and species in the first edition, he was disappointed that circumstances denied him the personal pleasure of turning the revered pages of the original edition. How delighted Coker, the avid collector of botanical volumes for the University, would have been could he have foreseen that, thirty-eight years after his futile search, the North Carolina Collection at his University would receive in 1981 the two precious volumes of the Catesby first edition through the bequest of Mangum and Josephine Weeks.

Coker's relentless pursuit of smilax specimens by making demands on his friends did not diminish their respect for him. Rather, these collaborators seemed more than willing to accept tasks they considered important. In 1943, the Charleston Museum made W. C. Coker "Honorary Curator." This was during the period when Robert Lunz, the hard-pressed smilax collector for Coker, was acting codirector of the museum.[28] Tarbox was his coworker in the field and at Brookgreen Gardens. And Allard's respect for his mentor and friend knew no bounds.

Indeed, the favors so willingly given by Coker's friends and colleagues were not one-sided. Coker often and willingly helped those whom he pressed into service as collectors in any enterprise: Alma Holland, later Mrs. Beers, who was relentlessly ordered to investigate the new site for a native *Rhododendron* for his article published in 1919, and who never achieved a higher rank on the faculty than assistant, became coauthor of two of Coker's books;[29] Frank Tarbox sought and received Coker's enthusiastic help in planting Brookgreen; and the Charleston Museum received his regular support. As for the poet Archibald Rutledge, perhaps Coker's friendship sufficed to assure his help in locating the southern boundary of venus flytrap near his home in McClellansville, South Carolina.

### William Chambers Coker, Detective

As reported in an article for the *Journal of the Elisha Mitchell Scientific Society,* Coker visited in 1910 the grave of Thomas Walter, author of *Flora Caroliniana* (1788). Coker considered Walter's garden to be one of the earliest botanical gardens in America and his book "one of the most complete and accurate works on American Botany of the 18th Century."[30] Coker narrated his adventure of traversing thick pinewoods and dense canebrakes where impenetrable bogs made this country "a paradise of wild things." Crossing the old Santee Canal on horseback, he searched for traces of Walter's eighteenth-century garden near Pineville and the plantation "Belle Isle" on the Santee River. A local inhabitant led him a half mile farther down the river. Reaching the grave, Coker copied the inscription on Walter's tomb. Though he could find no trace of Walter's renowned garden, he listed plants in the vicinity and collected many plants within the half mile of the grave site."[31]

In an article in the 1911 number of the *Elisha Mitchell Journal,* Coker reported his pilgrimage to the site of Andre Michaux's garden near Charleston, South Carolina. This garden was reported to be on land purchased by the French crown in 1786, before Michaux's arrival in Charleston as official botanist for King Louis XVI. Coker knew that the garden was located near Charleston and was supposed to be "400 or 500 yards from the railway line in full view of passing trains."[32] On board a Southern Railway train from Charleston, he noticed certain large trees visible from the Ten Mile Station. Though unable definitely to locate a garden in the overgrown area, Coker photographed an ancient magnolia and listed the plants still to be found there.[33]

The life and work of botanists, particularly those who had worked in the South, fascinated Coker. In the attempt to learn more about the con-

An old *Magnolia grandiflora* in the neglected late eighteenth-century garden of André Michaux, photographed by W. C. Coker around 1911. *Journal of the Elisha Mitchell Scientific Society* 27, pl. 10.

tribution of well-known southern botanists, Coker sought information by corresponding with descendants of the plant collectors whose writings he admired, among them Walter Ravenel and Curtis.[34] After gathering information beyond the published works of botanical pioneers, Coker initiated in the academic year 1921–22 a course in the history of botany to be always taught during winter term, when field trips were limited.[35] Persisting and growing in importance in the botany curriculum, the course was in some years offered as a "seminar" or as "studies in the history of botany" with a prerequisite of two courses in botany.[36] Coker particularly enjoyed this course, which he taught alone for ten years, after which Totten or Holland (Beers) assisted him. "History of Botany" appeared in the catalog each year for twenty-four years. Coker was noted in the catalog as teacher of this course until 1945, the year of his retirement, after which Couch assumed responsibility for it.

Particularly eager to know more about Dr. Gerald McCarthy, the deaf botanist trained at Gallaudet College in Washington, D.C., Coker made extensive inquiries about him in 1919 and 1920. Though McCarthy was considered the first official botanist of the state of North Carolina, his life and work were little known.[37]

Coker's fascination with plants led him not only to search the fields and forests of North and South Carolina as a collector but also to broaden the general knowledge of native plant distribution. He also followed all possible leads to learn more about the collections and gardens of his predecessors who had contributed to the knowledge of plants of the South. All of these interests would dovetail in his founding and developing the UNC Herbarium, a research collection of dried plants that today is the largest of its kind in the region.

CHAPTER FOUR

# Founder of the Herbarium

*Mr. Ashe's collection is . . . very large and
absolutely indispensable to any study of
southern flora. . . . To have this remarkable
collection go out of the state would be a calamity.*

—W. C. Coker to Frank P. Graham

UPON HIS ARRIVAL in Chapel Hill in 1902, Coker found no equipment
for use in teaching botany. On the fourth floor of New East, then the
location of the biology department, the working herbarium consisted of
"a few sheets, possibly a hundred or two mounted or unmounted plants
scattered about in corners and under tables, with little order."[1] William
Willard Ashe, whom Coker later came to know and admire, had collected
these plants during his botanical studies as an undergraduate at Chapel
Hill, from 1888 to 1891. In 1908, when the botany department was separated from the Department of Biology and moved to Davie Hall, Coker
founded, with Ashe's student collection and a few other specimens of
higher plants and mushrooms, what was to become the University Herbarium. The collection grew slowly during the teens and twenties. As his
letters attest, Coker spent much time teaching, performing administrative duties, writing, and editing the *Journal of the Elisha Mitchell Scientific
Society.** As faculty chairman of the Grounds and Buildings Committee, beginning in 1914, he oversaw campus landscaping. He had already planned

* The *Elisha Mitchell Journal* was founded at the University of North Carolina at Chapel
Hill in 1884 and named in honor of the distinguished geologist, mathematician, and

and directed the campus garden, known as "the Arboretum," since its beginning in 1903. Even so, by the early or mid-1930s Coker's collection of fungi for the Herbarium had grown to 20,000 specimens, while the number of woody plant specimens had increased through steady acquisition to 15,000.

Though the word "herbarium" is an ancient one that rolls easily from the tongue of botanists, many lay people find the term a total mystery. Even among the most ardent devotees of the University of North Carolina at Chapel Hill, most are unaware of the meaning of the term, or that the University's collection exists or that it is a national treasure. A definition of herbarium is thus in order: a collection of dried and pressed plants, preserved and mounted, labeled, dated, located, and organized for reference. Each specimen's sheet bears the name of the plant, its locality and habitat, the date of its collection, and the collector's name. Herbarium specimens are the only authentic source of identification of plants and of their present and past distribution.

William Coker was a devotee of herbaria. He made frequent use of preserved plant collections, both at home and abroad, to authenticate and complete his knowledge of the genera he was preparing for publication.[2] He visited the Farlow collection in the cryptogamic herbarium at Harvard, where also the herbarium of the Reverend Moses Ashley Curtis* was of particular interest to him; the Peck herbarium in Albany; the Schweinitz[3] herbarium in Philadelphia; the collection at the Lloyd Museum in Cincinnati; and the herbaria of the Bureau of Plant Industry and the Smithsonian Institution in Washington. The New York Botanical Garden reserved a research desk for Coker's use during his frequent visits there in the early decades of his career.

Coker traveled to Europe in the summer of 1921[4] expressly to "straighten

---

amateur botanist who served on the faculty of the University for almost forty years during the first half of the nineteenth century. Elisha Mitchell taught at Chapel Hill from 1818 until his death in a fall from a high peak in the North Carolina mountains in 1857. The highest mountain in North Carolina, indeed the highest east of the Mississippi, bears his name. The journal is now the official publication of the North Carolina Academy of Science. See Rogers McVaugh, Michael R. McVaugh, and Mary Ayers, *Chapel Hill and Elisha Mitchell the Botanist* (Chapel Hill: The Botanical Garden Foundation, 1996) 1, 3.

* Curtis (1808–72), renowned Carolina botanist, was born in Massachusetts and as an Episcopal clergyman served churches in both Society Hill, South Carolina (home of Coker's grandparents), and Hillsborough, North Carolina, near Chapel Hill, where he was buried. See Albert E. Sanders and William D. Anderson Jr., *Natural History Investigations in South Carolina from Colonial Times to the Present* (Columbia: U of South Carolina P, 1999) 59. A memorial plaque near the west transept of Saint Matthews Church in Hillsborough, where Curtis served, records the number of plants in his lifelong collection.

William Chambers Coker looking at a specimen of joe-pye-weed (*Eupatorium*) in the University's Herbarium. *Dr. Claire Freeman called my attention to this photograph from the "Hat's Off" section of the "Yackety Yack" of 1942, the UNC-CH student annual.*

out a few points in my Clavaria book, which is otherwise about ready to be published." In Stockholm, Lars Romell furnished him with Swedish material and access to the Bresadola herbarium there. Coker scheduled work in the Fries herbarium in Upsala, Sweden,[5] and the Persoon herbarium in Leiden in The Netherlands. At Kew, near London, he received access to material of both European and American origin. His exchange of specimens with mycologists, both professional and amateur, constantly broadened his understanding.[6]

Meanwhile, he worked constantly to expand the herbarium of his University. Wherever he went, he collected, pressed, and recorded specimens. Hartsville and its environs provided a rich source. His brothers James, David, and Charles were ever ready to provide him with specimens.[7] During his visits to the home of David Coker, his nieces watched him spread fresh plants between newspapers and clamp them with a wooden press. Notes on herbarium sheets reflect Coker's visits to the homes of family members no longer in Hartsville: to his sister Susan in Greenville or Caesar's Head, South Carolina; to the summer home of his brother James in Blowing Rock, North Carolina; to the home of his sister Jennie in Redding, Connecticut, or to his brother David's cottage at Myrtle Beach, South Carolina. The extreme heat of summer at Myrtle Beach did not prevent his daily quest for specimens of such unusual plants as the coastal chinquapin (*Castanea ashei*)[8] and the toothache tree (*Xanthoxylum clàva-hérculis*).[9]

Once, during a family day trip from Hartsville, Coker spotted a coveted

William Chambers Coker behind a rare toothache tree, *Xanthozylum Clàva-Hérculis*, to demonstrate its size. Photograph by H. R. Totten near Murrill's Inlet, South Carolina, July 15, 1932. *Southern Historical Collection, Wilson Library, University of North Carolina at Chapel Hill.*

plant in the swamp under a seven-mile raised highway bridge where drivers were forbidden to stop. He called a halt and asked the driver to continue to the end of the causeway and return for him presently. He then left the car and climbed over the guard rail into the boggy area. Upon the driver's return, his collecting foray still in process, he waved the car on. Indulgent relatives made several seven-mile trips before their passenger, unapologetic, his face scarcely visible behind his trophies, climbed back to the highway bridge from the rich swamp to resume his place in the car.[10] This incident illustrates how Coker's enthusiasm for collecting, transmitted to his colleagues and students, enriched the University's Herbarium.

### Acquisition of the Ashe Herbarium [11]

One of the most important professional accomplishments of W. C. Coker was his successful campaign to acquire the Ashe herbarium for the University of North Carolina.[12] The opportunity dramatically to increase the scope of the Herbarium at Chapel Hill arose without warning. On March 18, 1932, William Willard Ashe died unexpectedly.[13] Though a full-time employee of the United States Forest Service in Washington, D.C., Ashe was also a keen observer, collector, and annotator of the plants of his native state of North Carolina and of the South* The greater part of his lifelong collection was housed in his own home in Washington. A lesser part remained in a storehouse behind his family home, that of his father, Captain Samuel Ashe of Raleigh.[14] North Carolina botanists considered the Ashe herbarium to be uniquely valuable, indeed indispensable for any study of southern flora.[15] T. G. Harbison of Highlands, North Carolina, a close friend of Ashe who had tramped the southern woods and fields with him as fellow collector, and W. C. Coker, his friend and correspondent, began immediately to seek ways to acquire this valuable herbarium for the University of North Carolina at Chapel Hill. J. S. Holmes, North Carolina state forester and colleague of Ashe, was also eager for the acquisition.

Ashe left no written instructions for the disposition of his collection. Harbison's remembrance of conversations on collecting trips assured him that his friend had intended him to have carte blanche in the disposition of his considerable collection. Harbison favored the University at Chapel Hill, where vigorous botanical work was then in progress, as custodian of this fine collection of the North Carolina botanist.[16] The collection, however, belonged to Mrs. Ashe.

Harbison and Coker feared that acquisition of the collection from Mrs. Ashe would be next to impossible. North Carolina was in the depth of the Depression. Little money was available even for the barest necessities. Coker wrote to Harbison on April 2, 1932, "Past experience . . . makes me very doubtful of the money being found to pay for it. I have no idea what it is worth." Several other university herbaria were actively pursuing the acquisition of Ashe's vast collection: The National Arboretum wanted his types. In this context, "type" meant for Harbison and Coker the specimen Ashe used to illustrate his newly named species, the one to which all individual plants that bear that name must belong.[17] The urgency and difficulty of obtaining Ashe's precious collection consumed Coker and Harbison, even though each pessimistic letter included a discussion of other matters, such as the exchange of specimens for identification or the location in the wild of plants of prime interest to both botanists.[18]

---

* "In all Ashe published 510 new botanical names. Six trees and two shrubs of the Southeast, besides one widely distributed grass of the eastern states, perpetuate his name: *Castanea ashei, Crataegus ashei* Beadle, *Hicoria ashei* Sudworth, *Juniperus ashei* Buchholz, *Magnolia ashei* Weatherby, *Panicum ashei* Pearson, *Polycodium ashei* Harbison, *Quercus ashei* Trel, and *Schmaltzia ashei* Small." See the obituary of Ashe, written by W. C. Coker, J. S. Holmes, and C. F. Korstian of the North Carolina Academy of Science and published in the *Journal of the Elisha Mitchell Scientific Society* 48.1 (October 1932): 46.

In a letter dated May 9, 1932, Harbison stressed the need to raise funds to keep the Ashe herbarium in North Carolina. If Chapel Hill, his first choice, should fail to acquire this treasure, Harbison's next choices would be Duke University or State College in Raleigh. He reported to Coker that a gentleman in Florida might be persuaded to contribute toward purchasing the Ashe herbarium for Duke, Harbison's second choice as location for this valuable resource. Weighing other possibilities, Harbison mentioned the Arnold Arboretum in Boston. Having collected southern plants under its director, Dr. Charles Sprague Sargent, Harbison thought that Ashe's herbarium would be more useful there than at the New York Botanical Garden. In this letter to Coker, Harbison made the pessimistic comment that, though he was very much in favor of keeping the Ashe herbarium in North Carolina, the severe economic situation might indeed render this impossible. Nonetheless, he advised making "a strong effort in this direction."

In this letter to Coker less than two months after the death of Ashe, Harbison reported that Mrs. Ashe had invited him to Washington to arrange for the sale of her husband's herbarium. As Ashe's close friend and associate on collecting trips, Harbison was Mrs. Ashe's choice of an agent to sell the collection at the best possible price to whatever institution could pay for and protect it. Painfully ambivalent, Harbison was mindful of his duty to Mrs. Ashe, and yet he was eager to place the collection where it would be most useful to North Carolina botanists.

Writing to Coker on June 18, 1932, Harbison declared his decision to delay his trip to Washington until after August in order to gain a little more time to canvass southern institutions "in an attempt to keep the collection in the south." He informed Coker that a number of institutions were interested but that none seemed then to have the funds to acquire it. Even the Arnold Arboretum could purchase the treasure only "if the price were not too great."

On September 24, 1932, Coker wrote this urgent letter to his good friend, President Frank P. Graham of his University:

Dear President Graham,

I am [enclosing] a letter from State Forester Holmes, and a copy of a letter from him to Mr. Harbison and his reply to it. In spite of hard times I think you will agree with me that we should not fail in this emergency to make a strong effort to secure for the university this remarkable and unique collection of North Carolina and other southern plants.

In my opinion this is the most critical and important situation that has arisen in our department since I have been here. Mr. Ashe's collection is indeed remarkable; very large and absolutely indispensable to any study of southern flora. Mr. Ashe himself had a more intimate knowledge of southern flora, particularly in the trees, shrubs and certain grasses, than any other living man, and in his herbarium are found types of several hundred southern species named by him.

I do not know at present what this herbarium can be secured for but assume that $10,000 will be somewhere near the amount, possibly it can be had for less. Mrs. Ashe herself has no clear idea so far as to what it is worth, but all of us will know more after Mr. Harbison, Mr. Holmes, Mr. Totten and myself have had an opportunity to look it over.

You are aware that North Carolina may be considered at present the state in which the scientific and extensive study of southern botany is now concentrating. Work done here, at Duke and at State College is going on steadily all the time in the direction of a better knowledge of southern flora and to have this remarkable collection go out of the state would be a calamity.

After you have had an opportunity to think over this matter I would like to have an interview with you.

Sincerely yours,

This passionate letter had a profound effect on Dr. Graham, who himself was much concerned that his University excel in every way possible and who respected Coker as a scholar and a friend.

On October 10, 1932, J. S. Holmes, aware of the work of Ashe as a forester and botanist, wrote Dr. Graham telling him that he had given Coker names of a few possible contributors. He mentioned arranging for Coker to look over the Raleigh part of the herbarium. On the same day, Holmes wrote Mrs. Ashe quoting Dr. Graham as saying, "We want to do everything we can to get the Ashe collection and place it here in Chapel Hill." Holmes added that this was what, in his opinion, Mr. Ashe would have most desired. He also asked Mrs. Ashe for an estimate of the herbarium's value, assuring her of his assistance in the effort to find a market value for the herbarium of her husband.

Mrs. Ashe replied almost immediately, on October 12, that her own preference, and surely that of her husband, would be Chapel Hill as the location for the herbarium. She reported that one botanist had valued the approximately 30,000-sheet collection at twelve cents a sheet. She expressed doubts that true value could be obtained during the severe Depression. Dr. Holmes's reply on October 17 agreed with her suggestion that Dr. Coker or some other representative of the University call upon her in Washington to look over the portion of the herbarium in her home. In a letter to Coker written October 24, Holmes included a copy of Mrs. Ashe's last letter and suggested that Coker visit Mrs. Ashe on his next trip north.

Urgency was intensified on November 22 when Harbison reported to Coker in his small, neat hand that the University of Michigan would like to buy the Ashe herbarium "if it did not cost too much." On December 3, between paragraphs that mention his garden in Chapel Hill and the possibility of a station for *Magnolia fraseri* in Aiken, South Carolina, Coker informed Harbison that he now was in a position to make a direct offer for Ashe's herbarium, including both the Washington and Raleigh portions. Although President Graham had not yet found the money, Coker thought that an offer might encourage another extensive effort. Harbison replied two days later that in his judgement a definite price could not legally be offered until an appraisal had been made. The administrator of the estate wanted Harbison to see the collection. Harbison informed Coker that the Arnold Arboretum had funds to bid on Ashe's herbarium, that Ann Arbor had asked for a price, and that his own nephew, Thomas C. Harbison, forester at Pennsylvania State University, was seeking to obtain the coveted herbarium for the Pennsylvania State Forestry School. Most disquieting was the news that Penn State was attempting to obtain an appropriation for purchase of the Ashe herbarium by the direct intervention of Governor Gifford Pinchot, a formidable ally who had been chief of the U.S. Forest Service under President Theodore Roosevelt.

Harbison feared that the Ashe herbarium would serve merely as an exhibit at Ann Arbor. The Arnold Arboretum would indeed use it, but a location in Massachusetts would mean that he, Coker, and Totten would be "left out in the cold." Harbison continued in this urgent vein, "I think Ashe would agree with me that it should be where students of southern trees can have full and easy access to it. If we can secure it for the University, Chapel Hill will for years to come be the Mecca for students of southern botany as Charleston used to be when we had to know what Elliott was talking about before we could be sure we had something new." Harbison added a plaintive anecdote: "Years ago, I saved that valuable little collection [that of Stephen Elliott (1771–1830), specimens then in Charleston] from a dump heap where it was rotting, and now if I can do anything to save the Ashe collection for the south it will be a great satisfaction to me."[19] Referring again to his own painful position, he added, " I must not, however, be put in such a light in the sight of the public as to make me appear as working against the best interest of the Ashe estate. I am sure you understand my position."

Harbison ended his letter by giving Coker the name of a person in Augusta, Georgia, near Aiken, who could tell him more about where *Magnolia fraseri* could be found in the area. Thus, even in the intense struggle to obtain the Ashe herbarium for North Carolina, Harbison and Coker never ceased to communicate almost constantly about their first love, the identification and location of native living plants of the South, a passion that Ashe shared and that had inspired his relentless plant collecting. One senses Coker's urgency to obtain all possible information from Harbison, who so well knew the Ashe herbarium. Harbison's death, little more that three years later, was to deprive Coker of a second rich source of botanical information.

Harbison's letter to Coker from Highlands, dated December 14, 1932, bore the ominous news that Palmer of the Arnold Arboretum wished to meet Harbison in Washington to look over the Ashe collection. Palmer had suggested separating the herbarium into two sections: one section for the Arnold Arboretum with types[20] and a good representative of the trees first named by Ashe; the other for the University of North Carolina with cotypes and the bulk of the herbarium. Harbison considered this solution suitable only if he were allowed the selection of the cotypes. Typically, he had begun this urgent letter by telling Coker that he had never seen *Quercus borealis* (the mountain red oak) much below 4,000 feet in West Virginia or Tennessee, but that he and Totten had passed a few such trees on a hike to Clingman's Dome in North Carolina. Their common enthusiasm for plants in the wild was a safety valve for the two men, reducing the heat of the contest for the Ashe herbarium to a bearable temperature. Harbison concluded his letter with the discouraging news that he had little hope from any of those whom he considered able to donate the price of the Ashe herbarium. He mentioned a fine-spirited letter from the generous Mrs. Moses Cone of Blowing Rock, North Carolina, which included no assurance of a donation. Were they to be defeated by the rock-hard time between the election and the inauguration of Roosevelt, when the country was paralyzed and bank failures multiplied?

The first paragraph of Coker's reply to Harbison on December 17, 1932, dealt with the struggle for the Ashe herbarium. He wrote: "Your help in trying to get the Ashe herbarium for us is going about as I expected, of course. Money is hard or impossible to get now. If we could have a direct offer for the whole thing as I suggested once before, it might help a great deal. I wish you could secure this for us." The remainder of the letter dealt with the new book by B. W. Wells of State College, which he was sending to Harbison,[21] and with a proposal for a collecting trip "to settle a number of interesting matters that we both know need further investigation." He thought that Totten should join them.

Good news came a few days later and none too soon. On December 22, Coker asked Harbison to write him in care of his brother David in Hartsville that he would meet him in Washington the following week. He concluded with the good news, "President Graham and I succeeded yesterday in getting a gift from Mr. George Watts Hill of Durham of $3,000, plus a little extra for transportation and I hope that you and I can secure the Ashe herbarium from Mrs. Ashe for this price." Even then, this seemed too good to be true. "Needless to say," wrote Coker, "I am delighted at this opportunity and certainly hope that nothing will prevent the final securing of this herbarium for this University."

On January 5, 1933, Coker wrote Harbison in care of Mrs. Ashe in Washington, telling him that he had sent a personal check for the balance of $1,200 to complete the purchase of the herbarium. He asked Harbison to bring a full set of Ashe's papers, with all notebooks referring to his collections. The Raleigh material had already been brought to Chapel Hill, but Coker was delaying opening the boxes until the Washington portion, for which Harbison was responsible, arrived. The following day Coker again wrote to Harbison that he was to telegraph the Bank of Chapel Hill to confirm the payment by wire if the court did not accept his personal check.* Expecting to be reimbursed by the special gift for expenses, Coker sent another personal check to defray expenses. He commended Harbison for his decision to bring the herbarium to Chapel Hill by van, rather than to risk harm to the brittle precious materials in rail transportation.

Coker wrote J. S. Holmes on January 18 that the Ashe herbarium had arrived and thanked him for his help in getting the Raleigh section promptly moved. Whether Mrs. Ashe intended to give the University the books and pamphlets sent over with the herbarium was uncertain. Coker concluded his letter to Holmes by expressing his worry about the "inexplicable situation" of the disease killing sycamore trees. His letter concluded with an invitation for Holmes to look over the Ashe herbarium, even though it was in no condition to be viewed to advantage.

The same day, January 18, 1933, W. C. Coker's letter to his brother, David R. Coker in Hartsville, bore the good news of "this most valuable acquisition." The Ashe herbarium would be preserved in fourteen new

---

* During this dark period before Roosevelt's inauguration on March 3, 1933, banks were failing at such a rapid rate that all personal checks were suspect. Some recipients of checks raced to the bank to cash a check before the bank itself toppled. The Garden Club of North Carolina lost the sum of $587, which the women had collected for the support of the publication of B. W. Wells's *Natural Gardens of North Carolina*, in the failure in 1932 of the Commercial National Bank of High Point. See James R. Troyer, *Nature's Champion: B. W. Wells, Tar Heel Ecologist* (Chapel Hill: U of North Carolina P, 1993) 95.

cases in Davie Hall. He told his brother of his hope that Harbison would be employed to organize the new treasure.

On February 9, 1933, after a visit to Florida, Coker wrote Mrs. Ashe to thank her for her generosity in giving the department the literature left by her late husband.

## Organization and Growth of the University Herbarium

With the arrival of the Ashe herbarium, along with the literature donated by Mrs. Ashe, disorder again reigned in Davie Hall's botany department. Mrs. Ashe had estimated that there were 30,000 specimens in the collection. Many plants were simply in bundles. Sorting, poisoning,[22] mounting, labeling, and filing had still to be done. An herbarium curator was immediately needed. Dr. Harbison, Ashe's companion on collecting trips, was uniquely qualified to assume the responsibility of organizing the plants. The enemy again was hard times. The donor's funds in excess of the herbarium's cost could take care of Harbison's salary for only three months.

Needing to retain the services of Dr. Harbison while funds were being found to pay his salary, and knowing that his brother David needed help in planning and collecting plants for their prospective garden on the Bacot property in Hartsville, Dr. Coker suggested that Harbison go to Hartsville to help with the new garden.* David was to be responsible for his salary. The D. R. Cokers were delighted to have the advice and assistance of this distinguished botanist. There was some talk of Harbison's staying in the old house on the Bacot property on the outskirts of Hartsville, but the Cokers insisted on taking him into their home in town. On February 28, 1934, David wrote to his brother that Harbison was "a delightful guest, no bother." He remained in Hartsville from February 14 until April 18, 1934.[23] William Chambers Coker, intensely interested in the work at the Hartsville garden, wrote "Miss May," his sister-in-law, that he wanted to keep "current on the work under Harbison." Dr. Coker finally succeeded in persuading the University to hire Dr. Harbison as curator of the Herbarium, an agreement to be effective as of July 1, 1934.

Alterations at Davie Hall to house the new Herbarium were probably not yet completed; so Dr. Harbison did not begin work until late fall of 1934. The David Cokers requested Dr. Harbison's services again for a

---

* This is the property that William Chambers Coker had given to Mrs. D. R. Coker in March of 1932 so that she could preserve its natural flora and begin to create a botanical garden. This acreage with its unusual diversity of plants was to become Kalmia Gardens. See chapter 7 here for more about this gift.

short time in February 1935 to help obtain and plant azaleas for Kalmia Gardens, but he was not able to return to Hartsville.[24] At Chapel Hill, in October of 1935, Harbison, though in declining health, was again at work on the Ashe herbarium. He lived with the Tottens, devoting what attention he could to the Ashe materials, until his death on January 12, 1936.

Dr. Coker felt the loss of T. G. Harbison as a personal blow and as a tragedy for the botany department. Harbison seemed indispensable. He knew Ashe's collection intimately. In addition, he had become a virtual paleographer in his unique ability to read Ashe's cryptic shorthand field notes. Unfortunately, Harbison had scant time to work on the rich Ashe materials, obtained for the University with great effort and received in a state of disorder. Frustrating delays caused by the lack of space in Davie Hall, by the University's shortage of funds, and by Harbison's ill health had deprived Dr. Coker of all but about ten months of the much-needed help of this naturalist, who had become a living encyclopedia of Carolina mountain plants.

An intelligent and industrious graduate assistant, Laurie Stewart, had been working in the Chapel Hill Herbarium in September 1935. Coker made her supervisor of the Herbarium and director of four federal aid students until Harbison's arrival the following month. After Harbison's death early in 1936, Laurie Stewart kept the Herbarium operating as best she could. In her history of the Herbarium she described the bleak situation: "The little office where Dr. Harbison had been working on Ashe's *Crataegus* the previous fall was stacked with plants to be poisoned, mounted, sorted, labeled and filed away. Many of the new herbarium cases were stuffed with bundles of unmounted specimens in 'temporary' storage, there being no place to put them until they could be preserved and filed. Before long, the confusion would be compounded by the arrival of the additional 12,000 specimens of Harbison's own herbarium, recently purchased from his widow. Soon every case in the herbarium would be top-heavy with boxes and bundles of plants awaiting sorting and preparation for storage in those new cases."[25]

An incident occurred that prompted Miss Stewart to change the system of enumeration for herbarium cases. Apparently, Dr. Coker wanted to show a visitor a certain plant family in the herbarium. Going to case number XXVI, he found he had mistaken it for case XXIV. Embarrassed by his error, Dr. Coker began to mutter to himself. Stewart went over to help. Soon thereafter, Dr. Coker noted the cases bore familiar Arabic numerals.[26]

Stewart reported in 1940 a considerably larger and better-lighted work space for the forty-seven herbarium cases.[27] The tact and organizational ability of Laurie Stewart accomplished much for the Herbarium. And she was not without an ally. Stewart needed a plant drier, but Dr. Coker, remembering the blackened and moldy plants that came from the primitive drier she had improvised during her stay at Highlands in the summer of 1936, was reluctant to approve another. Dr. Totten, telling her that he would "work it out with Dr. Coker," quietly encouraged her to make an improved version.[28]

After Laurie Stewart received her master's degree under Coker in June of 1937, she became curator of the Herbarium. In 1937 and 1938, she busied herself learning the best methods of treating the treasures in her care. In November of 1936, she had visited the U.S. National Herbarium in Washington, the Gray Herbarium at Harvard, the New York Botanical Garden, and the Arnold Arboretum to learn about the techniques used in these distinguished, much older herbaria. She came away feeling that "in general we were quite up-to-date here in Chapel Hill."[29] By the end of the first year, Stewart and her student helpers poisoned and mounted the entire Harbison collection and were able to work in a well-organized office receptive to visiting botanists. The expansion of the new Herbarium was off to an auspicious start. Her first report in 1938 recorded the number of specimens to be 76,655. By 1942, the total number of specimens in the Herbarium numbered almost 100,000.[30]

In October of 1941 Laurie Stewart left her work at the Herbarium to marry Albert Radford, with whom she had worked since their meeting at the Highlands Laboratory in the summer of 1936. Upon his return from war service in 1946, Albert and Laurie Radford began collecting intensively for the Herbarium. Albert, taking on important work in addition to his many duties, became curator of the Herbarium after receiving his Ph.D. in 1948.

Though he kept on with his own research, Dr. Coker had begun to experience poor health in the late 1940s. His susceptibility to respiratory infections necessitated trips to Florida during the winter months. He continued, however, actively to acquire specimens and constantly to dream and work toward the growth of the University's usefulness in the collection of plants. He envisioned a separate garden to demonstrate the variety of trees and shrubs of the southeastern states, a dream not completed in his lifetime.[31] Coker and his associates collected and planted many varieties of native shrubs on the University property known as the Mason Farm.[32] As need for a University golf course in that area developed, Coker had many of the shrubs moved to the Arboretum on campus. W. C. Coker's dream of a demonstration shrub and tree garden was the harbinger of the present North Carolina Botanical Garden, a project founded and cultivated by his successors. Dr. Coker had mentioned Ritchie Bell as "a good one to head up the shrub farm." Dr. Bell was later to become founding director of the North Carolina Botanical Garden. This garden in a real sense has fulfilled Coker's dream of a shrub and tree collection, which Dr. Totten wanted to be called a botanical garden.[33]

In 1956, three years after Dr. Coker's death, three botanists at Chapel Hill, Albert Radford, Harry Ahles, and Ritchie Bell, initiated a systematic study of the flora of the Carolinas on a county-by-county basis. Intensive collecting resulted in the publication in 1964 of the *Manual of the Vascular Flora of the Carolinas*. Field research for this book resulted in the collection of another 200,000 specimens, the basic part of which were deposited in the University Herbarium.[34]

Dr. Jim Massey was curator of the Herbarium from 1970 until his retirement in 2000. During Dr. Massey's tenure, the poisoning and mounting of the many specimens collected for the *Manual* permitted their filing, their study at Chapel Hill, and their accessibility for exchange.[35] Massey created in the Herbarium at Coker Hall an agreeable work space for amateurs, as well as for professionals. He integrated the Herbarium into the educational program of students of biology. He organized Friends of the Herbarium, a community support group for the collection, and he also initiated the first herbarium newsletter in the United States.[36]

The University of North Carolina Herbarium, founded by W. C. Coker in 1908, houses, at the year 2001, 660,000 specimens.[37] It is the largest collection in the southeast. The National Science Foundation places the University of North Carolina Herbarium among 25 National Resource Centers and 105 National Resource Collections."[38] It now ranks fifth in size among all herbaria in the United States and third among university herbaria, after those at Michigan and Harvard. The Herbarium at Chapel Hill has exchanged more than 283,000 specimens with botanists all over the United States and in foreign nations, including China and the former U.S.S.R. An herbarium building, already designed for the North Carolina Botanical Garden, will eventually house the extensive collection at the University of North Carolina at Chapel Hill. Both the Herbarium and the University's botanical library are now, at this writing, crowded into inadequate spaces in Coker Hall.

From its small beginning in 1908, the University of North Carolina Herbarium in Chapel Hill has become a major resource for students of the earth's plant life. It is invaluable for scholars interested in the study of plants, especially plants of the southeastern United States. Indispensable to many research botanists, it has supported publication of twenty-four major works and hundreds of research papers. As a member of the International Plant Resources System, the Herbarium at Chapel Hill serves botanists worldwide.

Though including plants from many areas of the world, the University of North Carolina Herbarium at Chapel Hill is strongest in flora of the Carolinas and of the southeastern region of the United States. As such, it is of inestimable practical value for the state. It provides crucial information for planning North Carolina's future in the face of fast-paced development. North Carolina is home to 4,000 species of plants. The Herbarium documents their distribution among the 100 counties of the state. As a record of what we must preserve, the Herbarium at Chapel Hill is one of the University's most valuable archives.[39]

This collection owes much to the foresight and practical direction of the passionate botanist, William Chambers Coker, and to the enthusiasm and hard work of his distinguished associates to whom he passed the torch. The UNC Herbarium is as valuable to the University for the study of plants as are its famous Southern Historical Collection and its North Carolina Collection for the study of historic documents. This magnificent Herbarium is a testimony to the ability of William Chambers Coker to develop and then to institutionalize projects which he deemed of prime importance for the University. It is the result of his active implementation of a lifelong vision to record the plants of the Carolinas and the greater southern region.

# CHAPTER FIVE

# Champion of the Highlands Biological Station

*With his own personal direction and kindly leadership, Dr. W. C. Coker has been responsible for bringing the Highlands Museum and Biological Laboratory to a position of honored usefulness among the scientific institutions of the South.*

—Minutes, annual meeting of the board of directors, 1944

HIGHLANDS, NORTH CAROLINA, is a cool, high-elevation community of summer residents escaping summer heat in the lowland; its citizens have long supported their library, the first public library in North Carolina. The interest of these seasonal people and of other local citizens in indigenous culture and the natural world made Highlands a propitious place for the founding in 1927–28 of a museum for public education about the diverse life forms of the area. At an elevation of 4,000 feet and with rainfall of between 80 and 90 inches a year, the Highlands plateau is ringed by ancient, 5,000-foot mountains. Geological and climatic conditions have given this area what Ralph Sargent has called "possibly the most varied temperate mountain flora and fauna on the North American continent." He further described the Highlands area as "the biological crown of the southern Appalachians." The enormous number of life forms results from the fact that this plateau, with its rim of ancient mountains, guards an area little disturbed through the ages and thus has provided "a refugium for old forms and a cradle for new ones." As a result, the Highlands area of the Blue Ridge Mountains is alluring for working biologists.[1] William Coker called the Highlands area "probably the richest biological area east of the Mississippi."[2] Not only was the area particularly favorable for the collection of fleshy fungi, a passion of Coker's, but also for the study of trees and shrubs, another of his absorbing interests. Highlands was also an ideal location for Dr. Edwin Reinke of Vanderbilt University to study salamanders.

A lively community interest in the study of the natural sciences began in Highlands more than forty years before the first laboratory work in 1928. Thomas G. Harbison, who was to become Coker's close friend and botanical ally, founded in 1886 the Highlands Scientific Society, which met regularly to discuss natural phenomena of the area. From 1891 to 1895 the Southern Blue Ridge Horticultural Society, led by Henry O'Farrell and Charles Boynton, met for the discussion of various horticultural topics.[3]

William Chambers Coker was associated with the Highlands Museum and Biological Laboratory almost from its beginning. In August of 1927 the citizens of Highlands organized a museum to be housed in a one-room addition to the Hudson Library building on Main Street. W. M. Cleaveland, a local builder whose Indian artifacts were the museum's first collection, constructed this annex. Several local residents had added collections by the time the museum opened on July 4, 1928. Clark L. Foreman, the first president of the museum association, invited both Dr. Coker and Dr. Reinke to work at the museum that summer.[4]

The August 12, 1928, minutes of the museum's trustees record W. C. Coker's application earlier in the year for use of laboratory space that summer. At the same meeting, the director asked that water buckets be provided for Dr. Coker's use. Thereupon, the trustees moved that running water should be installed.[5] Thus began gradual improvement in the laboratory facilities.

In June of 1930, Dr. Reinke, who had become director of the Highlands Museum in 1929, and Clark Foreman, museum association president, along with Coker invited a number of biologists in the eastern United States to a conference to study the need for a research station at Highlands. Keenly aware of the extraordinarily rich biota of Highlands, the members of this conference approved on July 5, 1930, the incorporation of the Highlands Museum and Biological Laboratory. The original trustees, who were local residents, enlarged the board of the new museum and laboratory to include representatives of educational institutions. W. C. Coker had been one of the first trustees and became a member of the new board. Before the end of 1930, Cleaveland had begun work on the first laboratory building constructed at Highlands on Lake Ravenel. It was named the Weyman Laboratory in honor of Sam T. Weyman, the chief donor, an industrialist from Atlanta. Upon completion of the laboratory in July of 1931, a dozen investigators started work there, four of them from the University of North Carolina.[6] By the summer of 1932, both Coker and Reinke had established permanent summer homes in Highlands.[7] Coker's rustic cottage on Lower Lake Road, where he subsequently spent a part of every summer, still stands. Joe Webb built this house around 1922, the first of the famous log cabins he was to construct in the area.[8] Mrs. Florence Inman of Atlanta, Georgia, the present owner, cultivates a pleasant garden around the beautifully preserved historic house. Mrs. Inman has kept the old wooden "Coker" sign beside her own at the entry to the cottage on Lower Lake Road.

W. C. Coker drew a vivid picture of life during the first summer in his mountain cottage. In a penciled letter of August 5, 1932, to his brother David in Hartsville, he wrote, "Our cottage is very comfortable. . . . We have an excellent cook. Alma Holland and Nancy Eliason, a former student, are with me. . . . Our roof is full of flying squirrels. I saw 10 fly out a few nights ago and they run around the sitting room. We are taking a lot of hard mountain climbs and collecting. I killed a large rattler (10 rattles and a button) on one of our climbs last week. Am having Nancy prepare his skin."[9]

In August of 1933, at the invitation of the director and trustees of the Museum and Biological Laboratory, members of the Mycological Society of America met at Highlands for their first summer foray. Collecting was successful. On the evening of August 17, Professor Coker entertained the group at his home in what George W. Martin, vice president of the society, called "delightfully informal fashion."[10]

In 1933 and later, Coker and others made an effort to obtain support for the center, which was struggling under the constraints of the Depression. They tried unsuccessfully to interest both the General Education Board of the Rockefeller Foundation and the Carnegie Foundation in contributing to the center. As a concerned environmentalist interested in preserving the rich biological area for the American people and for biological research, Coker made a heroic effort to obtain federal protection for the "primeval forest" near Highlands. During the winter of 1936, he wrote to North Carolina senators Josiah W. Bailey and Robert Reynolds, as well as to several representatives in Congress, among them Zebulon Weaver and William B. Umstead, in an effort to obtain support for federal protection of the Highlands "primeval forest."[11] As late as 1939, Coker was still working to convince the chief United States forester, Mr. F. A. Silcox, of the importance of protecting this "wonderful forest of primeval growth." He even wrote to a member of Roosevelt's cabinet suggesting that he ask the President to himself say a word to Mr. Silcox.[12] Results of these persistent efforts were successful to a point but could not succeed, since he and State Forester Holmes could not persuade the

The Coker cottage at Highlands, North Carolina, now owned by Mrs. Florence Inman. *Photograph by William Joslin.*

owner to sell at a price the government would consider acceptable.[13] A letter to Coker of the previous December from Alvin G. Whitney of the New York State Museum mentioned "the importance of private ownership in the hands of a high-minded group with sufficient means to support and protect it." Whitney thus foresaw the key role that such groups as the Nature Conservancy would later assume in the protection of wild spaces in North Carolina.

The University of North Carolina became a member of the consortium of research institutions supporting the Highlands Museum and Biological Laboratory in 1933. That same year Coker became president of the corporation, a post he held until 1943.[14] For eight of these years, he was director of the laboratory as well.[15] He held the title of honorary president from the time of his resignation as active president until his death ten years later.[16]

A letter of Clark Foreman to Clark Williams dated September 12, 1936[17] reflects Coker's personal support as an on-site research botanist of the Highlands Laboratory. Foreman wrote, "Dr. W. C. Coker, Professor of Botany at the University of North Carolina and President of the Museum and Laboratory, is about the only man who has consistently, year after year, carried on planned research [at Highlands]."

Several persons associated with the University of North Carolina served as officers of the station, notably T. G. Harbison, J. N. Couch, H. R. Totten, Leland Shanor, and Albert E. Radford. Harbison, a long-time resident of Highlands, briefly became associated with the Department of Botany of the University of North Carolina in order to organize the Ashe herbarium, acquired by the University in late 1932.[18] Other research biologists at the Station from Chapel Hill or with connections to Chapel Hill include Alma Holland, Herbert Hechenbleikner, James Doubles, Laurie Stewart, Lytt Gardner, Frances Foust, Lane Barksdale, J. J. Valentine, Nelson Hairston, and Dr. Lindsey Olive, onetime chairman of the committee for the Station's botanical garden.[19]

The Highlands Biological Station (as it is known today) has produced important work in herpetology and in the study of fungi. The area around Highlands is well-known for its diversity of salamander species. The Station has sponsored several conferences on salamanders. "It has also been an important base of operations for mycologists, beginning with W. C. Coker in the 1920s. . . . In the summer of 2000 a course in fungi was taught for the first time in six years and drew a packed house."[20]

The present director of the Highlands Biological Station, Dr. Robert Wyatt, received his undergraduate degree in botany from the University of North Carolina at Chapel Hill. As a student, he regularly worked with Dr. Bell and Dr. Radford in Coker Hall and in the Coker Arboretum. He is the first botanist since Coker to become director of the Highlands Biological Station. His predecessor, Dr. Richard C. Bruce, a zoologist, published a number of studies on the salamanders of the area.

A grant in 1957 from the National Science Foundation made possible the construction of a much-needed new laboratory building at the Highlands Biological Station. Dedicated on June 19, 1958, this building bears the name of W. C. Coker, the Station's longtime president and director.[21] A wing added to the western end of the Coker Laboratory was ready for use in the summer of 1966. The main building and its addition now house laboratories, stockrooms, an herbarium of mountain plants, a library, and a computer facility. In 1967–68, a two-story administrative wing was added to the eastern end of the Coker Laboratory. It was dedicated in June of 1974 as the Thelma Howell Administration Building, in honor of the long-term resident administrator of the Station.

Miss Howell, with much help from Dr. H. R. Totten of Chapel Hill, was able to convince the North Carolina General Assembly in 1958 of the importance of the Station for higher education in the state of North Carolina. Representative James W. Raby, the Macon County legislator, introduced in the General Assembly a bill to provide annual funds for the Station.[22] This appropriation was the first step toward the station's be-

Robert Wyatt, director of the Highlands Biological Station at the entrance of the Coker Laboratory there. *Photograph by William Joslin.*

coming an integral part of the University of North Carolina system, a dream of W. C. Coker, who had suggested in 1941 that the laboratory be separated from the museum and offered to any or all participating institutions.[23] Dr. Reinke had thought it best that the University of North Carolina assume ownership. Coker later agreed with Reinke that this solution seemed the most reasonable one.[24] Thirty-five years after Coker's suggestion that the laboratory be associated with an academic institution, the state of North Carolina accepted the station as part of the university system. At the same time, the Highlands Biological Foundation was established to hold outlying property and endowment funds.[25]

By the year 2001, its seventy-fourth year, the Highlands Biological Station had the support of thirty-four southeastern universities and colleges. It is officially described as follows: "The Highlands Biological Station is a field station for biological research and education in the Southern Appalachian Mountains . . . an inter-institutional center of the University of North Carolina. It is administered by Western Carolina University, located in nearby Cullowhee, a constituent institution of the University of North Carolina. The facilities of the station are available for year-round use by qualified scientists who are engaged in research on the environments and biota of the Southern Appalachian region."[26]

William Chambers Coker was an enthusiastic supporter of the Highlands Biological Station from its beginning in 1928, when he requested modest working space with water buckets in the one-room Highlands Museum, itself an adjunct to the Hudson Public Library of Highlands. He remained a supporter until the end of his life. He contributed to the station, not only with work toward his own publications and with service as president and director, but also with financial help as needed. For example, in 1937 he gave to the Highlands Biological Station the land on which the Coker Laboratory was eventually built.[27] He contributed funds for the new museum building in 1939.[28] He helped substantially with the reconstruction of the dam on Lake Ravenel, breached in the torrential rains of 1940.[29] At the time of his retirement in 1944, he cancelled the remainder of the Station's debt to himself for the dam.[30]

Dr. Ralph Sargent, successor to Coker as president of the Highlands Biological Station, stated in a letter to W. C. Coker, written on September 17, 1944, after the annual meeting at which Dr. Coker's resignation was accepted, "I need scarcely say, what everybody knows, that the Highlands Laboratory, as a Southern mountain research station, is almost solely the creation of your own efforts."[31] Doubtless Coker would have disclaimed this credit. At the August 27 meeting of the board of trustees of the Highlands Biological Laboratory, when Coker was declared a "Patron" of the Museum and Laboratory, he protested the recommendation, stating that townspeople, summer residents, and visiting scientists had been equally devoted to the affairs of the laboratory. He was overruled with one dissenting vote, his own.[32]

Many publications over the years are based wholly or partially on research done at the Highlands Biological Station. Sargent gives a long list prepared by Richard C. Bruce and David Michener of those works published during the Station's first half century, "based in whole or in part on research conducted at the Highlands Biological Station."[33] A recent report records that from July 1999 to June 2000 the Station hosted more than fifty researchers representing more than twenty-five colleges, universities, and government agencies. Not included are researchers who commute from nearby institutions for senior investigations and working scholars who own cottages at Highlands.[34]

One may safely call William Chambers Coker one of the founders of the Highlands Biological Station at Highlands, North Carolina. He participated actively and regularly as a researcher, as an officer and director, and as a faithful supporter of the Station in every sense of the word. It is in no small measure due to the persistent efforts of W. C. Coker that the Highlands Biological Station, a part of his lasting legacy, has become a strong institution that continues to play an important role in biological research in southern Appalachia.

Student sketching with art class in the Arboretum.
*Photograph by Sandra Brooks-Mathers.*

Memorial gathering circle at entry to the Arboretum from the pergola
on Cameron Street. *Photograph by Sandra Brooks-Mathers.*

Great trees with sunlit distance. *Photograph by Sandra Brooks-Mathers.*

Fall scene among great trees. *Photograph by Sandra Brooks-Mathers.*

Perennial bed seen across the lawn. *Photograph by Diane Birkemo.*

Flowering cherries bordering McCorkle Place on campus. *Photographer unknown.*

Walter's pine, largest tree in the Arboretum. *Photograph by Sandra Brooks-Mathers.*

Fall foliage in the Arboretum. *Photograph by Sandra Brooks-Mathers.*

Small children enjoy the Arboretum. *Photographer unknown.*

Two students in a bed of Star of Bethlehem in the Arboretum.
*Photograph by Sandra Brooks-Mathers.*

Oak tree vista in the summertime. *Photograph by Sandra Brooks-Mathers.*

# *Founder of the Arboretum, Planner for the UNC Campus*

> *Among the memories of beneficial experiences
> during my pleasant years at Chapel Hill,
> one that comes to me often is your trans-
> formation of a half-drowned bit of meadow
> into the Arboretum.*
>
> —Frank McLean '05, to W. C. Coker

> *Other Universities have larger student bodies and bigger and finer buildings,
> but in Spring there are none, I know, so wonderful by half.*
>
> —Thomas Wolfe '20, to Lora French

DURING THE century of the Coker Arboretum's existence, the account of its origin and early years has assumed the quality of legend. It is the story of an idea that one man immediately seized and acted upon but only realized fully after years of persistent effort. It is the story of the love affair of a botanist with a piece of land and its possibilities. There are at least two different versions of essentially the same story. There are perhaps others.

Archibald Henderson, in his history of the UNC campus that was published for the University's sesquicentennial, recounts that on a morning in 1902, soon after William Chambers Coker arrived at the University, President Venable and the new botany professor were walking along the extension of Senior Walk. The president suggested to the new professor of botany that he take over the five-acre bog through which they were passing and try to make of it something pleasing and useful, perhaps a garden or an arboretum. President Venable mentioned to Coker that the original plan for the campus designated the southeastern area of the University's land for "ornamental grounds." Receiving this suggestion with enthusiasm, Coker began at once, with an appropriation of ten dollars and the help of one gardener, to drain and sub-tile the soggy acres. This was to be the work of several years.[1]

The second version of the Arboretum's origin is that of Curtis Brooks in his "Brief History of the Arboretum, 'The Coker Years.'"[2] Brooks recounts that Dr. Kemp Plummer Battle, former president of the University (1876–91) and then professor of history (1891–1907), walked daily to his office on campus from his home in Battle Grove, along a higher path through two wet pastures on either side of Raleigh Road. One day, President Venable and W. C. Coker joined him in this shortcut. Venable suggested that Coker might be interested in doing something to beautify the area. Presumably Battle, as one keen on the aesthetic appreciation of the campus surroundings, echoed this sentiment. Seizing upon the idea, Coker began immediately to draw up plans and to collect trees and shrubs for a garden. He first planted in 1903 the more elevated, drier northern border behind the Episcopal church, with a few American hollies, fringe trees, box elders, and later a Walter's pine he had collected near Charleston. Using some of his own funds for materials, he and one worker began to dig ditches and to lay subsurface tiles. The laborious work, though unflagging, went slowly. In 1907, four years after his suggestion that Coker make something of the area, President Venable supplied him with the services of a second worker and arranged for the Arboretum to receive a yearly budget of $350.* When drainage work was complete, the Arboretum contained over a mile of subsurface drainage tiles. By 1908, when the botany department occupied the newly built Davie Hall, Coker had improved the original central path over which Dr. Battle walked daily. In addition, he had designed a system of curving paths to make for easy accessibility and pleasing vistas. The Arboretum was attracting attention by 1909 when A. U. Caudell of the Bureau of Plant Industry in Washington wrote to ask Coker about his "trees, his orchards and the arboretum."[3] By the early teens planting was well established.

Fortunately, University planners have always considered this land unfit

---

* Throughout the years, the hard work of many local gardeners has created the beauty of the Arboretum we now enjoy. Among these persons, anonymous for the most part, are James Nunn, Sam Duberry, Sam Caldwell, Preach Perry, Louis Weaver, and Thomas Purefoy. Those who have kept the stone walls in repair include Dee Baldwin, members of the Blacknell family, William "Smitty" Smith, Jim Black, and Richard Johnson. See *A Backward Glance: Facts of Life in Chapel Hill* (Chapel Hill: Chapel Hill Bicentennial Commission, 1994) 34. Mrs. Frances Hargraves of the Chapel Hill Historical Commission and Charlotte Jones-Roe of the Botanical Garden have helped me to identify the names of some of these gardeners. We recognize with gratitude the work of them all, known and unknown.

President's Walk in the Coker Arboretum, 1909. *Southern Historical Collection, Wilson Library, University of North Carolina at Chapel Hill.*

President's Walk in the Coker Arboretum, ca. 1930s. *North Carolina Collection, Wilson Library, University of North Carolina at Chapel Hill.*

An early picture of the Arboretum, ca. 1910–1915, with Davie Hall at right rear. *Southern Historical Collection, Wilson Library, University of North Carolina at Chapel Hill.*

A later picture of the Arboretum, ca. 1920, of virtually the same scene as the previous one. *Southern Historical Collection, Wilson Library, University of North Carolina at Chapel Hill.*

for construction because of several underlying springs. The Arboretum is thus one of the areas of the central campus that has always been used for its original purpose. It remains today, a century after it began as the dream of one energetic man, in its original location, a place of quiet beauty in the historic heart of the University for the education and enjoyment of all.*

In response to the inquiry of a professor at Sweet Briar College in Virginia about a proposed arboretum there and its probable costs, Coker wrote on October 20, 1927, some practical advice related to his own campus project at Chapel Hill: "Our Arboretum of five acres costs about $2,000 a year now. Money has been spent all along, particularly in the early days, in underdraining, fertilizing and planting in addition to what one would call ordinary upkeep. Our so-called Arboretum budget is $3,000 a year but this includes propagating, planting and looking after all the shrubs on the campus, also the care of the greenhouse. Why not do as I did and begin on a small amount and gradually increase this amount as the arboretum grows. It will be easier to get money after results are visible." [4]

In October of 1920, in an effort to raise money for the beautification of the campus, Coker wrote to James Sprunt of Wilmington, "You perhaps know of our Arboretum, now including five acres, that I have been developing for ten years." This remark of the founder confirms the fact that the principal effort of the years before 1910 was to drain the terrain sufficiently to permit the growth throughout the area of the native trees and shrubs Coker wished to display in this campus garden. Apparently, Coker was actually planting the greater part of the area between 1910 and 1920. Thereafter, he continued constantly to increase the collection. Descriptions and photographs indicate that, though the Arboretum appeared in the early teens to be a vast lawn dotted with young trees, from the second decade of the twentieth century onward the campus garden was an increasingly lovely place. Toward the end of his active directorship, Coker wrote a detailed history and description of the Arboretum. This account appeared in 1949 as the greater part of chapter 25 in Dr. Archibald Henderson's sesquicentennial history of the Chapel Hill campus. [5]

---

* A straight path that traverses the Arboretum in east-west orientation, the President's Walk, follows one of two campus paths that existed in the early 1800s (Henderson, 121). I learned of the age of this historic path in a Christmas note, received in December 2000 from Dean Risa Palm of the College of Arts and Sciences. The Coker Arboretum, an integral part of the old central campus, today is bounded by Cameron Avenue on the south; the Morehead Planetarium, Howell Hall, and Davie Hall on the west; the Episcopal Chapel of the Cross and Spencer dormitory on the north; and Raleigh Road on the east.

Old section of Davie Hall from the Arboretum, early spring 2002. *Photograph by Sandra Brooks-Mathers.*

Meandering paths, the Arboretum, ca. 1930s. *Southern Historical Collection, Wilson Library, University of North Carolina at Chapel Hill.*

Three gardeners in the Arboretum stand before a *Cedrus deodara*, October 14, 1927. *North Carolina Collection, Wilson Library, University of North Carolina at Chapel Hill.*

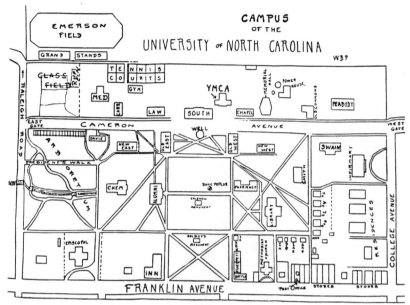

Plan of the campus of UNC showing the considerable size of the Arboretum
in relation to the size of the campus in 1917. *The Y.M.C.A. Handbook, 1917–18,
University of North Carolina.*

Though this five-acre campus garden has always been called the "Arboretum," Coker recognized the limitation of the site. Because the area remained too wet for many native plants, the founder preferred that it be considered an ornamental grounds for the instruction and enjoyment of all rather than a comprehensive collection of trees, which is an arboretum in the technical sense of the term.[6]

In 1918, Coker proposed to the president and trustees that the Arboretum be extended eastward and southward to the Forest Theatre to allow for successful growth of a wider variety of native trees. At the same time, he requested permission to plant a shrub garden behind Peabody Hall.[7] Though the proposed extension of the Arboretum across Raleigh Road was not approved, Coker was able to use land behind Peabody Hall. Here was a greenhouse, a gift of Mr. George Watts Hill of Durham, plus a nursery "for ornamental shrubs and trees both exotic and native." Coker used these plants, which he grew largely from cuttings and seedlings, to embellish the constantly expanding grounds of the University.[8]

By 1949, the southern border of the Arboretum contained a collection of some of the rarer native shrubs and a greenhouse, a gift of "the General Education Board."[9] On a long pergola bordering Cameron Avenue, Coker planted flowering vines, three varieties of wisteria, the native yel-

Plum tree in the Arboretum with Rebecca Ward, 1928. *Southern Historical Collection, Wilson Library, University of North Carolina at Chapel Hill.*

low jessamine (*Gelsemium sempervirens*), and Lady Banksia roses.[10] In his description of the Arboretum, Coker mentioned the rare Walter's pine (*Pinus glabra*), which he had brought from the Charleston area and planted many years before.* Today, this towering pine is the most imposing tree in the Arboretum. Its symbolism, to those who care for the Arboretum, is equivalent to that of the Davie Poplar in relation to the University. Both trees stand for a founder's vision, and both are living links to the past. Coker also noted another rare native plant, *Magnolia cordata*, considered a variety of the *Magnolia acuminata*,[11] and the river plum (*Prunus ameri-*

---

* This pine is named in honor of Thomas Walter, the eighteenth-century South Carolina botanist whose grave Coker found on the Santee River in 1910. See chapter 3 here.

*cana*), which blooms in late March. His favorite tree in the Arboretum was the Marshall's thorn (*Crataegus marshallii*), a red haw native to woodlands but seldom seen in gardens. Though this hawthorn was blown over after the Coker era, it still lives at the present writing, blooming and flourishing in a horizontal form.

From 1916 until after World War II, the Arboretum included a drug garden along its southern border. World War I, from 1914 to 1918 in Europe, had caused a scarcity of crude drugs in the United States, increasing the need for knowledge of botanical sources. Henry Roland Totten, later to receive his Ph.D. in botany under Coker, planted this garden as a graduate student in 1916 from seeds and roots of medicinal plants that Coker had requested and received from the United States Department of Agriculture. In 1917 Totten began his military training and subsequent service in France. Before returning home from World War I, Totten chose to prepare himself for continuing the drug garden by study at the School of Pharmacy in Paris in 1919.[12] For many years, Dr. Totten, using the Arboretum's drug garden as a laboratory, taught a course in pharmacological botany to students of pharmacy at Chapel Hill.[13] By 1925, 176 species of medicinal plants were growing there. Plants were laid out neatly in circular and rectangular beds.

The drug garden was identifiable in the Arboretum until after World War II, when emphasis shifted from living plants as pharmaceutical sources to synthetically manufactured drugs and to the study of the chemistry of drugs and their effects. Many experimental drug plants, originally planted near Mason Farm, were moved from their original location to the drug garden in the Arboretum to make room for the University golf course.[14] A section devoted to medicinal plants in the present North Carolina Botanical Garden south of campus continues the tradition started by Coker and Totten. There, in close proximity to the medicinal plants garden, is the Mercer Reeves Hubbard Herb Garden, which the Herb Society of America has chosen as the site for its National Rosemary Collection.[15]

Coker's list of plants in the Arboretum in about 1947, too long to be cited here, includes Japanese oak, viburnums, three varieties of buckeye (*Aesculus*), and the "tea olive" (*Osmanthus americana*), a large shrub that in September lends a haunting fragrance to old southern gardens. The *Lycoris*, or spider lily, grows and blooms well in the Arboretum, sending up naked flower stalks with spidery red flowers in September and providing a winter border of leaves after the blooms have faded.[16]

Though W. C. Coker retired as professor in 1944, he continued to serve as acting director of the Arboretum until 1947, when Roland Totten returned from his service in World War II to succeed him. The Arboretum

Vista in the Arboretum, with early spring flowering trees. *Southern Historical Collection, Wilson Library, University of North Carolina at Chapel Hill.*

finally received from the General Education Board of the Rockefeller Foundation its long-desired research and teaching greenhouse in 1942.[17] Native clethras and magnolias were planted in 1945 and the Japanese oak in 1946. In 1950, the Coker Arboretum, along with several other American botanical institutions, received a specimen of dawn redwood (*Metasequoia glyptostroboides*), grown at the Arnold Arboretum from seed sent from China. This tree, recently discovered and until then known only from fossil records, is now over fifty years old and is flourishing in the Coker Arboretum and in the gardens at Duke University.[18]

Other new and interesting plants came to the Arboretum in the 1950s. For example, one notices today along the northern border of President's Walk a variety of holly named "Pearle LeClair" [19] in honor of the first wife of Francis LeClair. A Belgian-born horticulturist, LeClair was landscape gardener and later University landscape architect during the years 1939 to 1957.[20] LeClair, a specialist in *Ilex*, planted hollies of many varieties on campus during his tenure at Chapel Hill. Robert B. House, in his hyperbolic enthusiasm, wrote that LeClair planted "enough hollies to stretch out seven miles if they were planted in a straight line." [21]

Maintaining the Arboretum became difficult with the tremendous growth of the University after World War II. Vandalism plagued the garden for a time. In 1956, Dr. C. Ritchie Bell became director of the Arboretum. He instituted a program of rejuvenation, thinning neglected clumps of plants to make room for the reintroduction of varieties previously rep-

Climbing roses at the entry to the Arboretum's pergola, 1935. *Southern Historical Collection, Wilson Library, University of North Carolina at Chapel Hill.*

resented in the Arboretum. Dr. Bell and others planted thousands of bulbs in the 1950s. For a decade, from the early 1960s to the early 1970s, the Arboretum was without a botanical director. It became the responsibility of the University's Buildings and Grounds Department.[22] Again the garden fell into disrepair. Dr. Lindsay Olive, the new director in 1973, reworked the plantings. Dr. Olive's tenure saw a renovated wisteria arbor along Cameron Avenue, an irrigation system covering the greater part of the Arboretum, the construction of barriers to prevent the encroachment of cars on the lawn areas, and a restoration of Elisha Mitchell's stone walls.[23]

Remember this walk in the Coker Arboretum on the campus of the University at Chapel Hill? You've probably walked down the path many times—and maybe you'd get a kick out of just seeing the picture. One fellow away overseas wrote: "I'd give a month's pay just to walk on the campus at Chapel Hill right now."

A scene in the Arboretum, from the *Alumni Review*, June 1944. The quotation in the top left corner reads: " Remember this walk in the Coker Arboretum in the campus of the University at Chapel Hill? You've probably walked down the path many times—and maybe you'd get a kick out of just seeing the picture. One fellow overseas wrote, 'I'd give a month's pay just to walk on the campus at Chapel Hill right now.'" *Courtesy of the photographic laboratory, Wilson Library, University of North Carolina at Chapel Hill.*

Francis LeClair, landscape gardener and later University landscape architect, 1939–57, with a blooming native fringe tree (*Chionanthus virginicus*) just off of the east end of Mason farm, ca. 1960s. *Southern Historical Collection, Wilson Library, University of North Carolina at Chapel Hill.*

In 1978, Dr. Clifford Parks became director of the Arboretum. He supervised the work of Paul Jones, who was the Arboretum's first curator (1978–84).[24] Janie Leonard Bryan served as Jones's assistant during the summer of 1978 and again in 1978–82. She is currently seed technician for the N.C. Botanical Garden. During this period, there was a great change in the Arboretum. Curtis Brooks describes the work of the Jones years: "Many dead, diseased and overgrown shrubs and trees, as well as volunteers of the common piedmont flora, were removed to make way for a substantial expansion in the plant collections. Notable among the many plants . . . [acquired] during this period are a large collection of slow-growing conifers, an extensive display of camellias and rhododendrons, and a diversity of deciduous flowering trees and shrubs, many of Asian origin."[25]

The succeeding hardworking curators of the Coker Arboretum who have helped to make it what it is today include Curtis Brooks (spring 1984–August 1986),[26] Diane Geitgey Birkemo (August 1986–October 1995), and Andrea Presler (1996–present). Dan Stern is her able assistant.[27]

Andrea Presler currently provides periodic guided tours for visitors to

European horsechestnut, planted by W. C. Coker, in
bloom on the campus of UNC. From Enid Matherly, "*How
to Know and Use the Trees*," Extension Bulletin *No. 3, 1924*.

the Arboretum, pointing out, for example, the blooming of the Japanese
apricot (*Prunus mume*) in early spring and the parsley haw or Marshall
thorn (*Craetagus marshallii*) in later spring. She calls attention to the
abundant scarlet fruit of the deciduous holly (*Ilex decidua*) and to other
remarkable hollies in winter and the overnight shedding of the golden
*Ginko biloba* leaves in fall. Ken Moore of the N. C. Botanical Garden also
provides guided tours of the old campus, following the example of the
late William Lanier Hunt, garden writer, former student of Coker, and
devotee of campus plantings.

The University at Chapel Hill has had its share of tragedies. The death, on September 13, 1912, of Isaac William Rand of Smithfield in a hazing incident opened up the investigation of other incidents and took its toll on President Venable, who later was to become Coker's father-in-law.[28] The murder of Suellen Evans of Mooresville, attacked by an unknown assailant in the Arboretum on July 30, 1965, stunned the whole state.[29] On May 12, 1996, five students perished in a fraternity house fire. Their tragic deaths inspired the Class of 1997, parents, and others to contribute as the senior class gift a lasting memorial in the Arboretum in honor of the students killed in the fire and three other classmates who did not survive to graduate. This memorial fund made possible the replacement of the old pergola along Cameron Avenue with a new arbor of durable native black locust. The project included a new southern entry into the Arboretum from the pergola to a stone courtyard, which serves as a gathering place for guided tours. Here a large stone leaf of tulip poplar (*Liriodendron tulipifera*), set into the flagstone floor at the center of the courtyard entry, points in the direction of the venerable Davie Poplar in McCorkle Place, several hundred yards distant from the Coker Arboretum in a northwesterly direction.[30] The new pergola and entry were dedicated on April 25, 1998.

The great age of the Davie Poplar, severely damaged in a storm in 1903, caused concern for Dr. Coker before 1917.* Symbol of the founding of the University in the eighteenth century, this tree that still stands on McCorkle Place has proven more durable than Coker hoped. In the future, after the inevitable demise of the historic Davie Poplar, the stone image of a poplar leaf at the entry of the Arboretum will direct the visitor's eye to the historic tree's former location.

The Coker Arboretum is not without its literary references. Thomas Wolfe, a 1920 graduate of the University, refers to the Arboretum in the journal he wrote on his first voyage to England. He records the antics of his Greek professor, "Bully Batson" (William Stanly Bernard):

> For years he was accustomed to go and come in a small, rubber-wheeled contrivance, scarcely eighteen inches in height, and wide enough for one seat before and one behind. In this fantastic little

---

* See photograph in Snider, 150. Coker wrote to President E. K. Graham on November 24, 1917: "I have been thinking about the Davie Poplar, and it seems to me that we should put the old tree in better shape, for looks' sake, if for no other reason. I think it will take about $60.00 to even up and put a tin cover over the hollow top of it and otherwise put it in good shape. Several years ago I recommended that the old tree be well braced in two directions and this was done. For some reason that I do not know the braces have been removed, and I think they should be put back by all means, as they may save the life of the tree for many years by preventing it from being blown down."

chariot he forsook highways, reeled down narrow paths, lurched precariously about and around the rooted great trees of the campus lawns, and might be seen in April, bursting suddenly from a flowering thicket in the Arboretum, coming out on two wheels under a gentle rain of lilac or dogwood bloom, spinning sideways down a steep embankment, and disappearing finally in a low thick curl of smoke and dust across the hill that leads to Durham.[31]

Students called Wolfe's Greek teacher, Professor Bernard, "Bully" behind his back. "In Wolfe's day, Bully's bachelor quarters (he later married) were at the intersection of Franklin and Raleigh streets, in a picturesque Italianate cottage, erected in antebellum times as a law office.[32] There he kept his Twomley, an automotive vehicle the shape of a bathtub with bicycle-size wire wheels. No more than two could ride in it, the driver and one passenger just behind. Wolfe called it a 'pussy cart.' . . . As Wolfe gradually got on more familiar terms with Bully, he was taken for rides, and it was a comical sight to see Wolfe 'squeezed into the back with his doubled-up knees projecting above the sides.'"[33]

Cutting across the Arboretum would have shortened the distance between Bernard's home and his office. Coker was concerned about automobiles on campus. One imagines the guardian of the Arboretum witnessing the wild ride of the Greek professor from a window in adjacent Davie Hall. Around this time, Coker recorded his own testimony to the misuse of the campus and the Arboretum by drivers of vehicles, though not in such lyrical prose as that of Thomas Wolfe. During Wolfe's junior year at Chapel Hill, on November 18, 1918, Coker wrote to Marvin H. Stacy, then dean of the faculty and briefly acting president of the University after the death of President Graham a few days previously. He suggested "the better policing of the Campus on Sundays and holidays, so that visiting autos may be kept in their proper places." He continued, "I notice considerable auto travel passing recently over foot paths through all parts of the campus and arboretum."[34]

In *Look Homeward, Angel*, Thomas Wolfe described the charm of the University in springtime. During the Wolfe years, the Arboretum was well-tended under the direction of Coker who, as faculty chairman of grounds and buildings, supervised campus maintenance and was active in its beautification. Wolfe wrote: "There was, God knows, seclusion enough for monastic scholarship, but the rare romantic quality of the atmosphere, the prodigal opulence of Springtime, thick with flowers and drenched in a fragrant warmth of green shimmering light, quenched pretty thoroughly any incipient rash of bookishness."[35]

In a letter to Lora French, written on May 17, 1920, near the end of his senior year, Wolfe paid tribute to the beauty of the campus at Chapel Hill: "Other Universities have larger student bodies and bigger buildings, but in Spring there are none, I know, so wonderful by half."[36]

Coker's work was not unnoticed elsewhere. W. B. McDougall, an instructor in botany at the University of Illinois, in a letter dated July 25, 1917,[37] requested material descriptive of the garden at Chapel Hill. Coker replied on July 28, 1917: "I may say that the main arboretum includes about five acres of ground, a part of which is laid off and included by hedges as a medicinal garden. We also have a special garden for the shrubs of North Carolina, including about two acres.[38] My principal object so far has been to get together the woody plants of North Carolina, and to make the place of pleasing appearance as an object lesson for parks and home grounds." In this letter, he implies that the Arboretum is part of the University's effort to give North Carolinians helpful examples for the beautification of homes or communities. Later, Coker and others introduced exotic shrubs and trees to supplement his collection of the native plants that could be grown in the Arboretum.

Supervision of the Arboretum was not without its headaches. On June 28, 1920, Coker reported to Professor N. W. Walker, director of the summer school, the evident lack of respect by summer students for the Arboretum. His own home garden nearby, at the corner of North and Boundary Streets, was another irresistible locale for romantic trysts:

> I should like to call your attention to the necessity of some kind of statement to the Summer School students in regard to the treatment of the Arboretum. The regular term students, who have learned something of University ways, very rarely cause any trouble in the Arboretum, but at the beginning and more or less through the entire Summer School, the temporary students cause considerable damage by picking almost any flowers, apparently, that they want. Much of this has been done already. I wish to make specific complaint especially against O. J. Sharp, who said he was a student. He went out of his way to pick one of the finest clusters of flowers in the Arboretum to give to a girl he was with, and when I spoke to him about it a few minutes later he would give no satisfactory explanation or excuse, simply saying that his name was O. J. Sharp and that I could take what action I pleased, or words with that meaning. This case is such an aggravated one that I think you should do something specific to discipline Mr. Sharp. If he is called in I do not think he will deny doing as I have stated above. It might be of interest to you to know that

almost any night at my place between eight and ten o'clock couples of the Summer School (apparently) may be found sitting on my piazza and on the bench in the summer house.

Always alert to practical ways to fill the needs of his campus garden, Coker wrote C. T. Woollen on June 21, 1920, "We are much in need of some good compost for use in the Arboretum and around several of the buildings, and I think there might be a good deal of it at the hog pen. Sam and Louis can get it raked up today, and I would like to know if the University truck cannot haul it for us tomorrow. Please O.K. this and tell Sam to go ahead getting it raked up. If the University truck cannot haul it, I will have it hauled with wagons."

Coker sought watchdog rules to protect the beauty of the campus. On June 1, 1920, he wrote four letters informing two professors, the YMCA secretary, and Mr. George Pickard, who evidently was responsible for campus men and wagons, that the Campus Committee, with the approval of the president, had decided to discontinue, at least for some time, all cutting of University shrubs and trees for use in decoration (presumably a custom at commencement).[39]

### Coker and Campus Planning

William Chambers Coker had early indirect contact with John Nolen, the distinguished city planner of Cambridge, Massachusetts and former student of Frederick Law Olmsted; Nolen was to design the UNC campus plan of 1919. In October of 1910, Coker received a letter from Charles N. Brown of the Madison [Wisconsin] Park and Pleasure Association, which informed him that Nolen was preparing a plan for Madison and that he had expressed the need for an arboretum for the city. Brown asked Coker for photographs of the Arboretum at Chapel Hill for possible publication in Nolen's report for Madison. In a second letter, written ten days later, Brown thanked Coker for the photographs and asked whether he should return them or turn them over to Mr. Nolen for his work.[40] In a letter dated February 7, 1915, Coker thanked Mr. John Nolen for his gift of a book to add to his Walt Whitman collection. In 1915, Nolen made a visit to Hartsville at the request of Coker's father and had sent James Coker a sketch of a plan for the campus of Coker College. On June 1, 1915, Nolen wrote to W. C. Coker informing him that he was sending his father his report on the campus at Coker College together with a little "sketch plan." He included in his letter to William a copy of this sketch.[41] Thus, when their correspondence about the UNC campus plan began in 1917, each man had an idea of the work and interests of the other, and each respected the other. They had indeed become friends.

For more than three years, from March 1917 to the end of July 1920, Coker corresponded with John Nolen about the campus plan and its implementation. On March 14, 1917, Coker wrote to Nolen asking for a definite plan for the grounds of the University in which he would locate all buildings, roads, and paths. On March 22, 1917, Nolen replied that he would let Coker know when he would be able to visit Chapel Hill. Coker quickly responded that he would look forward to an immediate visit.

On March 26, 1917, Nolen wrote Coker that he could not come so soon, that he would have to limit his stay to one day, and that he could not promise immediate planning. He sent Coker his formidable list of engagements and asked whether it would be better for the University of North Carolina to get someone else. Coker replied on March 30 that he had conferred with the president, who was willing to wait a month or five weeks, rather than choose someone else, even though President Graham was eager to have a plan before consulting with architects.

On April 14, 1917, W. C. Coker received a letter from President Graham enclosing a letter from Nolen requesting a topographical survey of the campus. Graham told Coker of the existence of such a survey but said that Mr. Woollen could not put his hand on it. He asked Coker to look into obtaining this survey. Nolen wrote Coker on April 21, 1917, of his plans to arrive in Greensboro "Monday morning at 7:05 and thence to make connection to Chapel Hill" and that he was looking forward to the visit.[42] The letters that have survived do not reveal whom Nolen visited or how many times he visited Chapel Hill. We surmise that Coker's usual hospitality was offered to his friend, for he often invited botanical associates, friends, relatives, and visitors to the University to stay in his pleasant home.

The war slowed down the rush toward getting the campus plan from Nolen. On February 16, 1918, Coker wrote to Nolen about his idea for extending the Arboretum across Raleigh Road and eastward as far as the Forest Theatre. In a letter of May 14, 1918 to H. R. Totten, then in the Army, Coker reported the completion of the plan.[43] He mentioned the prospect of a new quadrangle extending southward from South Building, with a gymnasium at the end and dormitories along the sides. On June 1, 1918, Coker asked Nolen if he could come to Chapel Hill to go over the whole situation. On June 5, 1918, Nolen expressed to Coker his disappointment that Phillips Hall, the new building between Memorial Hall and Peabody, had been placed too far forward.

On November 13, 1918, Coker informed Nolen about the death of President Edward Kidder Graham. He asked for the plan "before long." Nolen expressed his distress on hearing about the death of President Graham. He informed Coker that he was sending the results of the redrawing. On December 3, 1918, Coker received a letter from John Nolen asking for the return of the current tracing of the plans. A letter from Nolen to Coker dated January 23, 1919, inquired about progress the University had made on the plan, which he wished soon to finish. On January 25, 1919, Coker informed Nolen that he was sending him the original of his plan for the University grounds, together with certain suggested changes in one of the copies. He recommended that a student center, which was to be called Graham Memorial, be constructed "south of the Library at an equal distance from the Library and the path leading from the front of Alumni Hall," thus eliminating Building G. Coker suggested two treatments: that the two buildings be connected by a gallery or a loggia and that the paths at the north and south ends of the Alumni Building not be removed and that no professors' houses be located there. The women's buildings should be planned to accommodate more structures or a block of structures to be erected in time. His letter ended with the news of a second tragic death of the previous week, that of Dr. Marvin H. Stacy, dean of the College of Liberal Arts who was named chairman of the faculty on Graham's demise.[44] By the death of both of his administrative superiors, Coker inherited the entire responsibility for the University's evolving campus plan.

On February 6, 1919, Nolen informed Coker he had not yet had time to complete plans but that he would do so as soon as possible. Always eager to improve his own skills, Coker asked Nolen, in a letter dated February 12, 1919, to recommend a good work of reference on engineering as applied to landscape gardening and also on the technique of landscape design.

The Nolen plan, dated February 29, 1919, shows all of the existing buildings with fourteen new ones proposed. The plan retained the proposal for a great rectangular area south of South Building, which Coker had mentioned to Totten the previous May. This area would eventually be enclosed at the southern end by Louis Round Wilson Library, dedicated on October 19, 1929. Because of this creative proposal, we now enjoy Polk Place with its spacious lawns and double rows of willow oaks and white oaks now extending between the two distinguished buildings, South Building and Wilson Library, both monuments and guardians of the University's history.

John Nolen, in a letter dated March 6, 1919, informed Coker that he would receive under separate cover the revised plan for the University, showing redesigned paths and changes in the location of future buildings. He added that he had endeavored to embody in this final plan changes suggested in Coker's letter of January 29, the principal change being in the location of Graham Memorial. He wrote, "Perhaps this terminates this particular commission which I [Nolen] have enjoyed, notwithstanding the difficulties involved and delay resulting from war conditions." In his letter of March 11, 1919, Coker thanked Nolen for his patience and interest and invited him to visit Chapel Hill. No doubt impatient to have the completed, long-awaited document, he added that as of that date the plans had not yet arrived.

Over a year later, in a letter written March 31, 1920, Nolen asked Coker to send him a summary of what had been accomplished at Coker College and at the University of North Carolina since the respective campus plans had been prepared in 1915 and 1919. In his reply of April 16, 1920, Coker suggested that Nolen write Coker College about his sketch, as he had no memory of it, but that he himself had been in charge of landscaping the campus in Hartsville for the preceding four years. He reported that nothing had yet been done about the extension and arrangement of grounds at Chapel Hill except for the authorization of two new dormitories. He also reminded Nolen, "the plans finally adopted are different in many ways from the original ones prepared by you."

### Coker and the Implementation of the Nolen Plan

As faculty chairman of grounds and buildings, Dr. Coker was responsible for carrying out Nolen's plans, as modified by himself. He was concerned with employing the best architects possible for the buildings recommended in the campus plan.[45] Coker was in correspondence with the firm of McKim, Mead, and White in New York City during the spring of 1920. He informed the firm on April 14, 1920 that the trustees had finally approved their services as consulting architects and requested that they look over general ground plans and give suggestions as soon as possible. He added that a telegram had been sent the day before requesting an immediate visit by their representative. Though Coker had patiently to bide the time required for the legislative process and for the inevitable bureaucratic deliberations related to a state university, once the decisions were made, he wanted administrative action without delay.

President Chase informed Coker on June 26, 1920, that a representative of McKim, Mead, and White would be in Chapel Hill the following Monday "to try to bring matters to a head regarding our building operations."

On July 30, 1920, Coker wrote to a former student, J. A. McKay, in reply to a letter congratulating Coker on being awarded a Kenan professorship, that "I have just succeeded, after more than ten years' effort in getting the University to appoint a supervising architect. We have selected McKim, Mead and White of New York, one of the best firms in the country. We have also finally adopted a general plan for the University extension so as to avoid the haphazard confusion of the past. I believe our future growth will be along much better lines architecturally and in landscape design." [46]

From his correspondence of the years 1917 through 1920, it is evident that William Coker played an essential role in the formation of the "Nolen plan" and in the implementation of some of its recommendations. The two men, already friends before negotiations began, worked easily together in the exchange of ideas.

## William Coker's Role as Guardian of the Campus Landscape

In a letter to President H. W. Chase of April 30, 1920, Coker suggested that he be given authority to act in the interest of campus improvement. He wrote: "We are in a position to make our University and its grounds and forests the most notable in the South, and we only need the necessary machinery. It seems to me that the best way to spend money is in directions where immediate necessity is noticed and where immediate results can be seen with certainty and where the organization is already not only prepared but anxious to carry out the work." He again wrote to President Chase on June 9, 1920: "At present the University forests, including Battle's Park, are being badly neglected, with the result that there is much ill treatment and abuse. I suggest that the care of these adjacent woods be placed in the hands of the one you think fit, with power to keep them in reasonable order and to enforce certain regulations as to their use." [47]

In 1923, the Alumni Association of the District of Columbia, headed by Dr. Wade Anderson and Mr. Malcolm Weeks, mounted a campaign for the "Campus Beautiful Association," to raise money "to preserve and enhance the natural beauty of the Campus of our State University." Contributions were "to keep up the appearance of the rapidly expanding University grounds," as Coker, who served as treasurer of the beautification campaign, wrote to one contributing alumnus. This campaign prompted Frank McLean, a 1905 graduate from Maxton, North Carolina, then living in New York City, to write to Dr. Coker an eloquent tribute to the Arboretum: "Among the memories of beneficial experiences during my pleasant years at Chapel Hill, one that often comes to me, is your transformation of a bit of half-drowned meadow into the Arboretum. It was an inspiration of which I have many times been conscious and which has without doubt influenced me unconsciously even more. Now-a-days when there is in the South so much prosperity that is shared more and more by the man without a collar, it seems to me that there is no public service more needed or more intelligent than that of giving the students an environment of beauty." [48]

Archibald Henderson in his history of the campus had this to say about William Chambers Coker's Arboretum and other of his contributions:

> With hundreds of species of plants and trees artistically arranged for beauty and display, this is one of the most exquisite and harmonious small naturalistic gardens in the United States. This lovely spot, a haven of quiet, [is] like some 'garden close' of the Middle ages. . . .
>
> With an abiding sense of beauty and a disciplined passion for adornment, Dr. Coker has taken the Campus and indeed the entire Chapel Hill area for his province. As Chairman of the Committee on Grounds and Buildings for many years, he has been enabled to exercise a constructive and creative influence toward further beautification, through horticultural means and landscape architecture, of grounds and growths of rare natural beauty. Each new area, opened to Campus extension is treated with reverence for its qualities of natural beauty . . . and that beauty is suitably enhanced by artificial means.

Henderson predicted that future generations of students at Chapel Hill would have "a deep and abiding sense of gratitude to four great nature lovers and constructive artists who have left, for all to see, their testaments to beauty: Elisha Mitchell, David Lowrie Swain, Kemp Plummer Battle and William Chambers Coker."*

William Chambers Coker did not restrict his landscape planning to

---

* Henderson, 270. In 1838–39 Elisha Mitchell (1783–1857) personally demonstrated the technique of building stone walls, which he knew from his New England background, and supervised the construction of these walls around the UNC campus (Henderson, 126). David Lowrie Swain (1801–1868) considered campus beautification essential and called upon Mitchell's natural gift for landscape planning. He also hired English gardeners to improve the plantings of the north quadrangle, now known as McCorkle Place (Henderson, 110, 127). Kemp Plummer Battle (1831–1919) loved the forests adjacent to the University. He worked almost daily to make these woods more accessible and enjoyable for students and visitors (Henderson, 268).

the grounds of the University or even to his local community of Chapel Hill. As we shall see in the next chapter, he offered his advice and active help, insofar as he was able, to promote more pleasing private or public grounds in several areas, particularly for the beautification of public school grounds in North Carolina.

# CHAPTER SEVEN

# Landscape Designer, Extension Agent

WILLIAM CHAMBERS COKER, though untrained in landscape design, enjoyed using his natural talent for tasteful garden planning. As can be deduced from his plantings in the Arboretum, his was a combination of an aesthetic and an educational approach. His knowledge of how shrubs and trees fared under various conditions of sunshine, climate, and soil, and his instinct for pleasing combinations, led friends and representatives of public and private institutions to seek his advice. He planned the renovation of the old campus gardens at Sweet Briar College in Virginia.[1] He helped Professor Edward Caleb Coker with plans for the campus of the University of South Carolina.[2] He designed the campus of the newly founded Coker College in his hometown of Hartsville, South Carolina. Although John Nolen visited Hartsville at the invitation of W. C. Coker's father and subsequently sent both the elder Coker and William what he called a "sketch plan" for the Coker College campus, his plan was not used. It was William Chambers Coker who planned and supervised the campus landscaping between the years 1916 and 1920.[3] He also advised the grounds chairman at Coker College about how and where to order the plants he recommended.[4] He landscaped gardens in response to requests from family and friends, sometimes also ordering for them the plants he suggested.[5]

W. C. Coker had an important part in the planning and creation of Kalmia Gardens in Hartsville, the special project of his brother David and his sister-in-law May Coker. During the Depression, in March of 1932, he arranged for his nephew Richard to acquire for Mrs. David R. Coker (May) some acres west of Hartsville then called the "Bacot Place," the lo-

Outdoor theatre on the Coker College campus, designed by W. C. Coker. *University of North Carolina* Extension Bulletin *No. 3, 1924.*

cation of the historic house of Thomas Hart. As a youth Will had tramped this area, also called "laurel land," admiring the unusual growth of mountain plants on the north-facing slope of Black Creek. This quiet gift by Coker to an avid gardener, who herself later added several acres to the tract, became Kalmia Gardens. The planning and planting of this extensive garden became May Coker's life work. W. C. Coker, the donor, followed the growth of this garden in Hartsville with great interest. On occasion he offered advice about the acquisition of plants and suggested professional helpers.[6] Over thirty years after Dr. Coker's gift in 1932, Mrs. Coker presented Kalmia Gardens to Coker College in memory of her husband, David Robert Coker, to be used as a botanical garden for the college and a place for community enjoyment. Thus, W. C. Coker played a major role in extending opportunities for the appreciation of native and exotic plants in the community of his birth.

Brookgreen Gardens, which Archer and Anna Hyatt Huntington founded in the early 1930s near Pawley's Island, South Carolina, benefited from William Coker's help in plant propagation and acquisition. Coker was a trustee of Brookgreen for three years, from 1944 until 1947. He had, however, a congenial association with Frank Tarbox, chief horticulturist for the gardens, from the time of their meeting in April of 1937. The two botanists regularly exchanged seedlings, small trees, and flowers and fruit of various plants and shrubs.[7] Coker advised Tarbox on what would grow in his coastal-area garden and where to place plants for best effect. In

Kalmia Gardens, the "Purple Garden." *Courtesy of the D. R. Coker family.*

exchange, Tarbox kept an eye out for peculiar low country native plants that Coker needed for his research.[8] This assistance was especially helpful to Coker during World War II, when automobile trips were restricted by gasoline rationing.

Coker did not confine his interests to gardens and landscaping in Hartsville, Chapel Hill, and Brookgreen. Requests for advice on planting came to him from a variety of sources. In response to requests for his help, he planned the landscaping for industrial buildings in Durham, North Carolina; sent seedlings to the wife of a former student, a professor at Elon College, to improve the appearance of the railroad station at Elon; sketched a design for a cemetery in Asheville, North Carolina; and planned the grounds for the town library at Edenton, North Carolina.[9] His interest in the beautification of public and private buildings across the state of North Carolina seemingly had few limits.

## Coker as Extension Agent for North Carolina

In addition to his volunteer work in landscaping, Coker served the state of North Carolina in an official capacity, as an extension agent for the University. President Edward Kidder Graham conceived of the University of North Carolina as a means to enhance the quality of life for North Carolinians. During Graham's administration the University was open to any reasonable request for help where need or where opportunity for constructive service was demonstrated. President Graham "sought to make the University's boundaries coterminous with those of the state."[10] Agreeing with this concept of the service role of the state university, Coker offered the state one of his own talents. He began work in 1916 under Louis Round Wilson, director of the University's Bureau of Extension from its organization in September of 1913.[11] He assumed responsibility for the designing and planting of school grounds in response to written requests for help from North Carolina public school officials or even from community leaders.

There is reason to believe that Coker had given some thought toward improving the looks of the many depressingly bare school yards in North Carolina long before he began actively to serve as school landscaper for the extension program. It is possible, even likely, that he himself suggested to President Edward Kidder Graham and Dr. Wilson that he serve the state in this capacity. Certainly he accepted the assignment with enthusiasm. In 1906 Coker read and filed for future reference an article on children's gardens in *Plant World*, making notes and writing the comments "good" and "very good" beside two of the design plans.[12] In January 1908, he requested and received from President L. C. Lord of Eastern Illinois State Normal School an issue of his school garden bulletin.[13] In a letter to Mrs. Lula Martin McIver of Greensboro, Coker volunteered in May of 1909 to lecture to members of the Woman's Association for the Betterment of Public School Houses. After some delay because of a lost letter, Mrs. McIver expressed her pleasure that he would give the lectures.[14]

From early 1916 until the early 1920s Coker cheerfully responded to appeals from many North Carolina communities. For a time, he was also offering extension lectures in various towns of the state as a representative of the University, work he was obliged to discontinue as requests for school grounds landscaping multiplied.[15] He published in 1921 and 1924 two extension bulletins showing designs for school grounds and offering practical advice for their implementation.[16]

Requests for school landscaping during the years 1916–20 came from principals, school board members, or other persons interested in community improvement in many North Carolina towns: Burgaw, Chapel Hill, China Grove, Durham, Fayetteville, Goldsboro, Graham, Greensboro, Jackson Springs, Jamestown, Kinston, La Grange, Mt. Gilead, Nashville, New Bern, Selma, Spencer, Spring Hope, Stantonberg, Wilkesboro, Wilmington, and perhaps others.[17] Coker accepted the work without question,

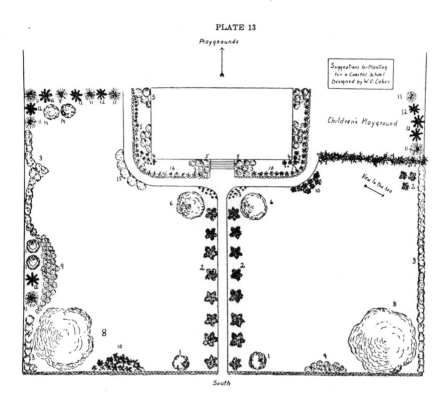

PLATE 13

## KEY OF PLAN FOR A SCHOOL NEAR THE COAST

(PLATE 13)

1. AMERICAN OLIVE, *Osmanthus americanus.*
2. PALMETTO, *Sabal Palmetto.*
3. YOPON, *Ilex vomitoria.*
4. SPANISH BAYONET, *Yucca gloriosa.*
5. WAX MYRTLE, *Myrica cerifera.*
6. MAGNOLIA, *Magnolia grandiflora.*
7. CHEROKEE ROSE HEDGE ON FENCE, *Rosa laevigata.*
8. LIVE OAK, *Quercus virginiana.*
9. OLEANDER, *Nerium oleander.*
10. *Tamarix gallica.*
11. CEDAR, *Juniperus virginiana.*
12. PINE, *Pinus Taeda.*
13. YUCCA, *Yucca aloifolia.*
14. DOGWOOD, *Cornus florida.*
15. TEA, *Camellia thea.*
16. LUPINE, *Lupinus perennis.*

W. C. Coker's schoolground design and suggested plants for a coastal North Carolina school, with key. Extension Bulletin, *Special Series No. 1, 1921, pl. 13.*

asking first for a topographical map of the grounds in order to adapt his plan to all physical and man-made features of the area. As compensation, he asked only for travel expenses to and from the site.[18]

Because of the heavy demand for school grounds designs, Coker hired at his own expense an assistant, Miss Eleanor Hoffman, who traveled to North Carolina schools and helped in drafting plans.[19] These plans, accompanied by suggestions for appropriate plants, illustrate Coker's skill in recommending shrubs and trees suitable for the climate and soil conditions in question. Nor did he forget to adapt his plans to the children. He allowed for playgrounds. At one point he recommended that nut trees and conifers not be planted along the street beside the school grounds, as children would be tempted to throw nuts and cones at passing cars. In the early twenties, cars were still a novelty in rural North Carolina and attracted not only attention but also occasional mischief.

Such was the influence of E. K. Graham's belief that the University should be a resource for fulfilling the state's needs that long after his premature death in the influenza epidemic of 1918 his ideals continued to direct the outreach of the University. This public service legacy of Graham and Coker is reflected today in the University's many services for the state, among which is the educational program at the North Carolina Botanical Garden.

### Landscape Design and Practical Advice for Chapel Hill

Dr. Coker designed the grounds of a church in Chapel Hill as a memorial to the wife of the donor, James Sprunt of Wilmington. Mr. Sprunt wrote him as follows on October 29, 1920: "I must take this opportunity to thank you again for your personal and professional kindness and assistance in the laying out of the grounds of the Presbyterian Church which Dr. Moss [the pastor] tells me is about ready for use. I am told that your excellent taste is manifest in the arrangement of the shrubbery and grounds."[20]

As director of the Arboretum, Coker shared with members of the community plants he propagated for use in the Arboretum and on the greater campus. On November 18, 1918, he sent the following notice to the *Chapel Hill News*: "Ornamental shrubs free of charge to persons in Chapel Hill interested in improving their yards by planting of shrubs. The University Arboretum will give free of charge a considerable collection of desirable plants. The plants will be for distribution to first comers—to those who apply at the Arboretum Saturday morning of this week. W. C. Coker, Director."[21]

Dr. Coker took great interest in the appearance of the village of Chapel

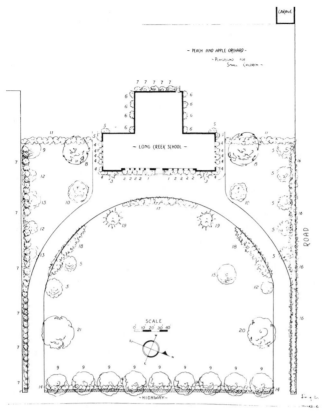

**KEY TO PLANTINGS FOR LONG CREEK SCHOOL**

(PLATE 19)

1. Sweet Breath of Spring.
2. Van Houtte's Spirea (between basement windows).
3. Red Cedar (Clipped back until high enough to go over path).
4. Golden Bell (*Forsythia Fortunei*).
5. Dogwood.
6. Ibota Privet.
7. Lilacs.
8. Pin Oak.
9. Sugar Maple.
10. Magnolia.
11. Hedge of McCartney Roses.
12. Crepe Myrtle.
13. Mimosa.
14. Hedge of Japanese Barberry.
15. American Linden.
16. Bridal Wreath (*Spirea prunifolia*).
17. Thunberg's Spirea.
18. McCartney Roses.
19. Deodara Cedar.
20. Willow Oak.
21. White Oak

W. C. Coker's schoolground design and suggested plants for a piedmont North Carolina school, with key. Extension Bulletin, *Special Series No. 1, 1921, pl. 19.*

Hill. He advised about the treatment of the space around the Chapel Hill post office and made other suggestions for town plantings. Presumably because of his interest in the beauty of the town, he was appointed in 1916 to fill a vacancy on the Chapel Hill Board of Aldermen created by the resignation of Dr. C. L. Raper.[22]

Coker was jealously protective of the plantings in the main commercial district of Chapel Hill. At any threat to a tree, he sprang into action. George Watts Hill, who served with Coker on the Chapel Hill Tree Committee, had given a merchant permission to severely trim a tree to make more visible a sign at his shop. Coker wrote to Hill: "Some of these merchants over here, very shortsightedly in my opinion, seem to want to turn Chapel Hill into a place like any other little one-horse town trying to ape a city, and I think that you, and I, and everybody else who is familiar with the character and flavor and distinctive qualities of Chapel Hill ought to do everything we can to prevent such a thing."[23]

That Coker had early assumed the role of garden advisor for Chapel Hill is apparent in three 1915 articles on lawns and grasses published in the *Elisha Mitchell Journal.* These articles responded to frequent inquiries about lawns from friends and neighbors.[24] In "Winter Grasses of Chapel Hill," Coker wrote with his characteristic mix of scientific description, personal observation, and practical advice. He listed and commented upon fifteen different grasses suitable for lawns. He described their particular features, their physical requirements, and where they could be seen growing in Chapel Hill. For example, a certain sedge is "an abundant constituent of the deeply shaded lawn at the old Holmes Place," and a delicate variety of sheep fescue "is plentiful in the lawn east of Dr. Manning's on Rosemary Street."[25] In "The Lawn Problem of the South," Coker's complicated instructions for a perfect lawn ended with practical advice to his fellow citizens of Chapel Hill, most of whom, as faculty members, had a modest income: "If you can afford a lawn mower you will have the one thing needful to improve the appearance of your home 100 per cent. Simply get rid of the sprouts and big weeds and run the mower over whatever comes."

This second article concludes with a lyrical description of how attention to one's home grounds has power to improve the community and one's personal outlook. "When you see this great improvement already made," he wrote, "you will not be quite satisfied until you take down that old sagging fence and plant a hedge in its place. Then, as you grow in grace and in love of beauty, you will add shrubs to the corners and about the house, shape up the walks and keep them hoed, and screen the unsightly places with evergreen privets or mock orange (*Prunus carolinianus*, not osage orange, which is not evergreen).[26] There will be joy in your heart at these transformations, and when, some day, you realize that the neighbors are trying to follow your example your full reward will appear."[27]

Coker reveals himself unmistakably in these lighthearted articles on lawns to be the practical neighbor, eager to enhance the appearance of his community. As editor of the *Mitchell Journal*, he here grants himself a practical voice, which he might have denied in another aspirant for publication in this scholarly journal.

### Efforts to Improve the University Campus

As chairman of the Grounds and Buildings Committee for the University from 1913 until the 1940s, Coker was tireless in his efforts to improve the appearance of the campus. In a letter to Mr. James Sprunt of Wilmington, dated October 28, 1920,[28] Coker suggested that he, along with four others, give $1,000 each for improving University grounds:

> I wish to say, in asking you to join us in getting up a fund for beautifying our surroundings here, that I am not doing so in a perfunctory way or as a matter of duty, but because I am most deeply interested in seeing this University make itself the most beautiful in the southern states, as an object lesson in methods by which we may introduce into our country the civilizing influences of more attractive surroundings. I have been working constantly to this end since my connection with the University, and I have little doubt that I can get you and others to help.
>
> As to just how this money should be spent should be clearly designated now. On a rolling hillside adjoining Battle's Park the University is just completing a new development of ten cottages for the faculty, and I am now trying to get the grounds of this development in shape, as the building operations are about over. We have done a lot of grading and have started some road-making, and the place is beginning to show what could be done with it if the necessary funds were at hand. It seems to me that we could scarcely do anything better now than to assist the University in making this place really worthy as an object lesson in such suburban development. The present poverty-stricken condition of the University will prevent them from doing anything more than the most necessary grading, and I propose that we devote half of the money that we get together, that is $2500, to the purpose of finishing and beautifying this place in best style.
>
> You perhaps know of our Arboretum, now including five acres,

that I have been developing for ten years. I am sending you some prints of views in it that will give you an idea of the place. I propose that the remaining $2500 be spent in carrying out this work. Our plans include also the building of a few drives through the fine woods belonging to the University, leading to the main points of interest, as Piney Prospect, Meeting of the Waters, Judges' Spring, etc., but this is not our immediate concern.[29]

### The Home Garden of William Chambers Coker, a Demonstration Garden at Chapel Hill

In June of 1906, Coker bought sixty acres from H. H. Williams and his wife, as well as five acres from O. B. Tenney and his wife, land then north of the village of Chapel Hill.[30] His purchase included both cleared and forested land and several tenant houses. The most striking feature of the property was a large outcrop of boulders. Here in 1908, on a hill behind these boulders, he built his house, which he called "The Rocks." He planted the surrounding land with orchards and gardens that featured native plants and certain exotic trees. Rhodes Markham of Chapel Hill was the gardener at his home place for many years. At the time of Coker's death, the land around his house comprised fifty acres. In the final settlement of his estate in 1954, Mrs. Coker was assigned the home place, where she remained until her death in 1983. Thereafter a portion of the property was subdivided and sold for homesites. The present owners who reside at "The Rocks," Dr. Walter Woodrow Burns and Mrs. Mary Jane Burns, have restored the home and tend there a lovely garden which includes the monumental *Cedar of Lebanon* of which Dr. Coker was so proud. They have taken pains to preserve some of Dr. Coker's original plantings and landscaping features, such as rock walls and stone pathways. They also help maintain, along with the North Carolina Botanical Garden staff, a small public park among the boulders on North Street, which Mrs. Preston Fox has provided with a permanent endowment fund in memory of her aunt, Mrs. Coker.[31] Dr. James Peacock and Mrs. Florence Peacock care for the area that includes what was once the formal garden, bounded on the north by a tall hedge of American holly (*Ilex opaca*), which Coker brought in from the woods nearly ninety years ago as an experiment to determine their sex and to demonstrate the great variety exhibited in native plants.[32]

During his lifetime and beyond, Coker's lovely garden at "The Rocks" was a demonstration of what could be done with property in town, just as the Arboretum was a demonstration of a campus garden. He collected

"The Rocks," home of W. C. Coker in 1923. The child is Coit Coker, later a marine biologist and hero of D day in World War II. He was the son of W. C. Coker's first Cousin, Professor R. E. Coker, who arrived in Chapel Hill in 1922 as professor of zoology. *Southern Historical Collection, Wilson Library, University of North Carolina at Chapel Hill.*

W. C. Coker garden at "The Rocks" with house in the background, 1923. The child on the wall is Coit Coker. *Southern Historical Collection, Wilson Library, University of North Carolina at Chapel Hill.*

Former Coker home as restored by Dr. and Mrs. Woodrow Burns, late 1990s. *Photograph courtesy of Mrs. Mary Jane Burns.*

Scene of conifers on the W. C. Coker home grounds after a snowfall. Photo taken by Dr. J. K. Small of the New York Botanical Garden and first published in the *Bulletin of the New York Botanical Garden* 31, pl. 251, 1920. *Southern Historical Collection, Wilson Library, University of North Carolina at Chapel Hill.*

Four visitors in the long pergola at the W. C. Coker home garden, probably in the 1920s. *Southern Historical Collection, Wilson Library, University of North Carolina at Chapel Hill.*

Six-stemmed yucca in bloom in the W. C. Coker garden, date unknown. *Southern Historical Collection, Wilson Library, University of North Carolina at Chapel Hill.*

for the acres surrounding his home a variety of both native and exotic shrubs and trees, arranging them to border a lawn to the west of his modified prairie-style house. A vine-covered pergola, similar to that beside the Coker Arboretum on campus, led to garden "rooms" more formal in design than the plantings around the large lawn area. He welcomed friends and visitors, whom he and Mrs. Coker often entertained, and took pleasure in showing them around his unusual collection of shrubs and trees. In thanking him for his hospitality during the 1917 meeting of the North Carolina Academy of Science at Chapel Hill, Professor John F. Lanneau of Wake Forest called his home and grounds "unique in elegance and beauty."* One of his former students remembered the comment of an English visitor whom he accompanied on a stroll around the Coker home—that she had at last seen in the United States an English garden.[33] The same student, Paul Titman, described his "valedictory" with Dr. Coker in his garden as follows: "He spent a long time talking about this plant and that, looking at the Bhutan pine, looking at the fern-leaf beech, looking at the grove of pawpaws, on through the garden and all around. I somehow think that we both may well have known that this was our last trip through this magic garden."[34]

The garden at "The Rocks" was an extension of Coker's teaching and an example to all visitors of his vast knowledge of plants, his taste in garden design, and his love for the beauty of nature. It reflected, as did his extension work for school grounds and as do the Arboretum and the central campus of the University, William Chambers Coker's taste and skill in the practical aesthetic use of plants.

* Letter of Lanneau, dated May 2, 1917. SHC. John Francis Lanneau, of a Charleston Huguenot family, was professor of physics and chemistry at Furman University 1858–61. He served the Confederacy for four years in Hampton's cavalry and after the war taught physics and applied math at Wake Forest College in 1890. Later, he was professor of applied math and astronomy at Wake Forest from 1899 to 1921. See his obituary and photograph in *JEMSS* 37:1–2 (1921): 17–18.

# CHAPTER EIGHT

# The Teacher and His Students

*A walk from his home through the village and Arboretum might suggest an entirely new topic for that day's discussion and the assistant who had prepared the classroom demonstration could never foretell when a complete change might be made at the last minute before the lecture began. But the lecture fitted the time and caught the students' interest.*

—John N. Couch and Velma Matthews,
"William Chambers Coker," *Mycologia*

*I shall always remember your kindness to me.
You have been more of a teacher than you know.*

—H. R. Totten to W. C. Coker

AS A TEACHER, William Coker aimed primarily to capture the interest of his students. While requiring careful laboratory work, he also encouraged students to remain close to the natural world. He assigned field trips to study living plants and often went along. When asked about his methods, he warned against too great a reliance on book knowledge as opposed to practical experience. However, he also considered laboratory work and the written word of capital importance. In 1917 he wrote to a professor at Urbana, Illinois, "I must say that I have never yet seen any substitute for hard work in the laboratory and with books which is the real foundation on which the broader and more genial views that come from outdoor observation should be based." [1]

In response to an inquiry about introductory botany, Coker described his own course. He said, in part, "I think it is suicidal to try to introduce, even by name, several of the so-called fields in botanical science. . . . An intelligent student is quick to perceive any superficial handling of a subject and at once reacts unfavorably when he meets it. The student should be presented with realities that he can see, feel and measure without too much theory or correction of error. The bearing of these facts on theory, on the phenomena of life in general, and on human life is a matter for the teacher to present in the best way that his training and skill will allow." Coker then outlined his introductory botany course of the preceding quarter, going into detail about each subject, from the lower to the higher plants. He mentioned his requirement of a paper on a lower plant, plus a thesis—a study on an entire higher plant, roots, leaves, flowers, fruits— "as far as this can be done. . . . Plates from several of the best theses are selected and framed for hanging in the laboratory, in this way collecting a series of plates which would do credit to the best textbooks." He required two field trips to introduce students to about fifty species of trees that they should later identify from specimens on examination.

His introductory botany course had to serve students with diverse requirements: Some in first-year botany were working toward the bachelor's degree in liberal arts, many were expecting to go into medicine, and others were farmers wanting to learn more about plant growth. As North Carolina during the early decades of the twentieth century was largely a farming state, Coker thought it important that a summer course be offered for the study of all growth stages of farm crops.[2]

In response to an inquiry from Professor A. Caswell Ellis of the University of Texas at Austin in November 1919 as to the proper services of a professor and the characteristics of good teaching, Dr. Coker described his own work at mid-career:

> A professor should teach up to about ten hours a week if necessary, not more if he can help it, less if possible. He should keep at research all the time. He should do as much committee work as he can do well and that he finds diverting. I see no reason why he should not edit a journal in his subject or advise on student editorial boards. Such drudgery as editing the catalogue, reading large numbers of quiz papers, essays, etc., should not be forced on professors. As to matters of administration, they are, in my experience, assumed by professors in exact proportion to their desires, far too many taking the easy way and drifting into this kind of work.
>
> So large and obscure a subject [as to what characterizes good teaching] will probably be answered differently every year, even by the same professor. My present impressions are as follows: Present certain subjects rather fully. Do not teach too many topics until one is

thoroughly understood: i.e., relate all amplifications, extensions and exceptions to well understood nuclei. Do not be too bookish. Let the class collect their own material and take frequent excursions. Have objects in classroom and show and compare them frequently. Give much time to laboratory work. Have a good text and make the class study it. Touch on the human side often in the lectures. Refer class to good books for voluntary reading and try to get the student interested in the great men of the subject. Send students to the board often to organize their ideas, the end being to select and present subjects in such a way as permanently to influence and improve the thought and action of the student. Such is "practical" in the true sense.[3]

Coker was not bound to his syllabus. Two of his former students described his flexible method as follows: "A walk from his home through the village and Arboretum might suggest an entirely new topic for that day's discussion, and the assistant who had prepared the classroom demonstrations could never foretell when a complete change might have to be made in the last few minutes before the lecture began. But the lecture fitted the time and caught the student's interest, and so often contained fresh observations on living plants and was so graphically presented that the student was encouraged to find out something new about plants for himself."[4]

## Cooperative Research with Students and Colleagues

Coker recognized in his publications the work of undergraduates who made notable observations. As an example, Adrian Couch in 1930 confirmed a description of pores in the cross walls of a species of *Saprolegniaceae*. The young Couch received credit for his keen eye in Coker's article on a new species of fungus in the *Mitchell Journal* of that year.[5]

Coker was quick to perceive exceptional interest in botany and to encourage it. A bright student in his introductory course of 1939 was surprised by Coker's proposal that he become an undergraduate teaching assistant in botany. Though then intending to specialize in art, the student accepted Coker's challenge and ended up choosing botany as his prime interest.[6] Paul Titman's story of his transition from art to botany casts light on Coker's approach to teaching:

I went to Chapel Hill and began to work in the atelier of the art department under the leadership of Russell Train Smith, the chairman of the department. One of the requirements was that one take a general botany course in the education spread. I was happy to do so

Paul Wilson Titman, as
a UNC graduate in 1941.
*General Alumni Association
records office, University of
North Carolina at Chapel Hill.*

because of my lifelong interest in plants.* . . . I still remember the
first day of class, when a very distinguished, slim, elegant, white-
haired, rosy-faced man, with a clipped white moustache came in and
in a deep South Carolina accent introduced himself as William
Chambers Coker. He also announced that this was the last time he
was going to be teaching the introductory course. I soon found that
his approach to life mirrored my own aspirations. To put it into
terms of my present perspective, he regarded botany as an humanity,
a manifestation in the middle of all kinds of thought processes. . . .
He [the art advisor] went to Boston and I went to Botany.

W. C. Coker was keenly interested in his graduate students' research
in progress. He worked alongside them as a colleague in the laboratory.
He used class notes of some of his best students in subsequent teaching.
Among Coker's papers is a notebook of Orren W. Hyman[7] from his Botany 3 class during the spring of 1911. Coker also kept a botany notebook
of W. L. Goldston Jr., a geology student of Dr. Collier Cobb who received
both the A.B. and M.A. from Chapel Hill the same year (1916).[8] Another

* Titman here refers to earlier remarks on how he had learned about plants as a child
during long walks with his grandmother, who with a native herbal remedy had saved his
life in his infancy during a siege of typhoid fever. Titman's audiotaped reminiscences,
December 1998, author's collection.

notebook of Introductory Botany, dated February, 1915, has the following inscription and is signed in Totten's hand: "The main part of this note-book were notes taken in Dr. Coker's class by H. R. Totten, who had charge of the laboratory, but who attended the lectures again (third time) to better tie the laboratory work with the lectures. The notebook was given to Dr. Coker by Totten and Dr. Coker used it for the basis of his course later."[9]

Totten immediately became his professor's respected colleague. Three years after Totten's graduation from the University in 1913, Coker published his first tree book, *The Trees of North Carolina*, with Totten as co-author.

Dr. Coker conducted many research projects in conjunction with his students, both graduate and undergraduate. These research projects often resulted in publications of joint authorship. In the bibliography of *Blastocladiales, Monoblepharidales, and Saprolegniales* in *North American Flora*, in which Coker has more entries that any other writer, eight student co-authors are listed: Louise Wilson, W. Alexander, Herman Harrison Braxton, J. N. Couch (three articles), F. A. Grant, James Vernon Harvey, Velma Matthews, and J. D. Pemberton.[10] Coker also credited as coauthor in one or more of his other publications thirteen former students or colleagues.[11]

### Coker's Influence on Selected Students; Their Contributions to Biological Science and Teaching

His lasting influence on the life of students began early in Coker's career. For example, Harry Ardell Allard, Class of 1904, from Oxford, Massachusetts, was already well advanced in university studies when Coker arrived at Chapel Hill in 1902. Just eight years Coker's junior, Allard stayed at the University after his graduation, until the spring of 1905, as his botany teacher's laboratory assistant.[12] He was helpful in early plantings at the new Arboretum. For Coker's article on conifer spores published in the *Botanical Gazette* in 1904, Allard outlined in ink twenty-four drawings.[13] He received credit for his contribution in a note. The mutual esteem between Coker and Allard was lifelong. Allard's admiration for Coker is poignantly apparent in his answer to an alumni office questionnaire of January 23, 1923: Noting the names and birth dates of his three living children, Allard added that his second son, William Coker, had died at the age of three months.

One of Coker's first serious students, notable even among his stellar students of the following years, Allard made important contributions to the biological science of his day. He served for forty years at the federal

Bureau of Plant Industry, researching ways to improve agricultural plants. But his work and writings were not limited to this useful work in Washington. He studied and wrote about nature as an independent scholar. He is perhaps best known as codiscoverer of floral photoperiodism and as author of many published studies of insect sounds and habits. With his keen eye and ear for events in the natural world, Allard resembled his botany professor at Chapel Hill.

Two years after H. A. Allard's retirement from government service in 1946, the University of North Carolina conferred upon him the honorary degree of Doctor of Science.* Both Allard and Coker were doubtless much gratified that this son of the University, who had received no further degree since graduation, was so honored.

Two remarkable men came under William Coker's direct influence during the second decade of his tenure at Chapel Hill—Henry Roland Totten[14] and John Nathaniel Couch.[15] The lives of these two men were so closely allied with that of Coker and so nearly contemporaneous as almost to be mentioned in the same breath. Totten's description in 1938 of his own career during the twenty-five years after his graduation not only includes a tribute to Coker but also curiously reflects in himself some of his teacher's personality traits, just as Coker's own skills and aspirations bear the imprint of Duncan Starr Johnson, his botany professor at John's Hopkins.[16] Totten wrote:

> Most of my life since graduation in 1913 has been spent in Chapel Hill, and teaching botany. It has been a busy, but pleasant life in a beautiful and freedom-loving community. Quite a few students have passed through my classes during these twenty-five years. . . . I have tried to teach my students something of the methods and exactness

* The tribute to Allard reads as follows: "Harry Ardell Allard, B.S., North Carolina, 1905, where he worked his way through college; W. C. Coker's first laboratory assistant; helped in the first plantings in the now famous Coker Arboretum, worked under Dr. H. J. Webber in the U.S. Department of Agriculture in the breeding of corn, cotton, pineapples, sorghum and broom corn hybrids. His original research in tobacco induced a new outlook and interest in virus diseases; his work in the photoperiodic behavior of plants became a turning point in the understanding of plant growth and reproduction as influenced by seasons and climatic conditions in all parts of the world; author of 300 scientific papers largely dealing with the ecology of plants and the stridulations of insects. A naturalist who loves plants, people and life anywhere; one of those pioneer scientists, who in the quiet of nature's laboratories, blazes modestly the trails which become the highways of the world's life. By vote of the Faculty and Trustees of the University of North Carolina we confer upon you the degree of Doctor of Science." (Text courtesy of Ms. Tracy Chrismon of the General Alumni Association records office, Chapel Hill.)

Harry Ardell Allard receiving the honorary degree at the UNC Commencement of 1948. W. C. Coker straightens his hood. *General Alumni Association records office, University of North Carolina at Chapel Hill.*

of the laboratory, the scientific method and spirit. Probably the most successful and enjoyable work for them and for me has been the field work—long hikes over the Orange County hills (one of the most favorable places in the world to study plants) where we really learn plants and each other. Much of the vacations, when finances would permit, I have spent in field work through the southeastern states collecting and studying the distribution and habits of the woody plants of this region. I have had a small part in bringing together the very valuable collections of plants in the Herbarium of the University of North Carolina. . . . In this study of the woody plants I have had the special privilege of working with W. C. Coker and T. G. Harbison—two men with seeing eyes and the true scientific spirit. I have learned from both of them.[17]

Henry Roland Totten,
as a UNC graduate in 1913.
*General Alumni Association
records office, University of
North Carolina at Chapel Hill.*

Professor Coker's enthusiasm as a teacher dramatically influenced the career choice of John Nathaniel Couch. In 1917, during his junior year at Trinity College, now Duke University, Couch, strongly attracted to the biological sciences, decided to prepare for medical school. Needing to meet course requirements in botany, he transferred to Chapel Hill to complete his undergraduate requirements. While studying mycology with W. C. Coker, he made an abrupt change in his life plan and decided to become a botanist. Though his studies were interrupted by military service in World War I, he returned to Chapel Hill in 1919, eventually to be awarded all three degrees in botany under Dr. Coker.*

With his teacher's encouragement, John Couch attached his interests quickly and permanently to mycological research. And he readily acknowledged Coker's seminal inspiration. In February of 1925 Coker wrote to Dr. Charles B. Davenport of Cold Spring Harbor, recommending Couch

---

* Among his other honors, Dr. Couch became a member of the National Academy of Sciences in 1943. In 1981, he received the distinguished mycology award from the Mycology Society of America, one of the first to be given this honor. See William R. Burk and Charles E. Bland, "John Nathaniel Couch 1896–1986," *Mycologia* 81.2 (1989): 181–89; and L. Shanor, "The Career of John Nathaniel Couch," 1–7 in W. J. Koch, ed., *Mycological Studies Honoring John N. Couch, JEMSS* 84.1 (1968):1–280. The author remembers that in the spring of 1945 Dr. Couch sought her out to tell her of his debt to Dr. Coker as an inspiring teacher.

John Nathaniel Couch, botanizing in Iowa in 1923. *Courtesy of Mrs. Sally Couch Vilas.*

for a national research fellowship in Dr. Blakeslee's laboratory. Over a year later, as his fellowship there drew to a close, Couch wrote Coker about his experience, "I have enjoyed my work here this year, but I must say that by far the pleasantest and best working moments I have ever spent have been in your laboratory."[18]

Many of Coker's students have made significant contributions to biological research and teaching. Such stars among his students as Clyde Ritchie Bell, Lindsay Shepard Olive, and Albert and Laurie Radford, who have themselves contributed much to the University of North Carolina at Chapel Hill, are not treated in this context, for they appear elsewhere in these essays. Here we select for mention four students among many others who flourished under Coker's influence. (1) A Tarboro graduate of 1910, Orren Williams Hyman received the master's degree in biology at Chapel Hill in 1911 and the Ph.D. in biology from Princeton in 1921.[19] He served as vice president of the University of Tennessee Medical Units from 1925 to 1961. On a questionnaire for the alumni office he listed W. C. Coker among his favorite professors, the other two being "Froggy" Wilson and Horace Williams.* (2) Called "Virginia's Mr. Moss" or "Dr. Pat" at Hollins College, where he taught for many years, Paul M. Patterson received the M.A. in botany from UNC in 1927. He was president (1951–52)

---

* See General Alumni Association records office, Chapel Hill. On January 8, 1944, Hyman wrote to Coker to express his appreciation for his teacher's influence on his life: "My interest in Botany was not deep, as you knew at the time and may possibly recall—but your friendship and the field work I did under your stimulation did give me a little knowledge

Velma Matthews, holding flowers of *Zenobia cassine-folia* on a field trip, May 1926. *Southern Historical Collection, Wilson Library, University of North Carolina at Chapel Hill.*

and later, for eight years, secretary of the Virginia Academy of Science. He also served as president of the American Bryological Society.[20] (3) Velma Matthews received both the M.A. (1927) and Ph.D. (1930) in botany at Chapel Hill. The first woman president of the South Carolina Academy of Science (1946–47), she was from 1934 until her death in 1958 professor of botany at Coker College.[21] (4) As professor of biology at Coker College, Dr. Budd E. Smith made an exhaustive inventory of the plants of Darlington County, South Carolina. He is included in a list of distinguished botanists from the seventeenth century through the first half of the twentieth century who "immeasurably increased" knowledge of the flora of the Carolinas.[22] He later became president of Wingate College in North Carolina.[23]

of a wide variety of living plants. As a consequence I was quite an active botanizer for some twenty years and get quite a bit of pleasure that way yet. Believe it or not I can recognize most of the flowering plants of these parts at sight. My knowledge of the mushrooms is feeble and insecure and my dim light in the field of molds flickers out with *Thranstotheca clavata*. "Life has been—and is—fun to me and my interest in Botany has added materially to my fun. Mille mercis for nudging me into the field!" SHC.

William Chambers Coker, about the age of fifty. *From* Mycologia, *Vol. 46, 1954.*

### Coker's Personal Interest in His Students

W. C. Coker's interest in his students went beyond his encouragement of their studies and writings. He showed for many of them both the respect of a colleague and the concern of a family member. Considering them to be endowed with great potential, he expected his students to work diligently. Advanced students often became his assistants. Being Coker's student meant not only hard work but also an invitation to share botanical adventures, an invitation to whimsy. One of his former students characterized Coker as "a right jolly old elf." This descriptive term for Clement Moore's plump and indulgent Christmas midnight visitor applied to the twinkle in his teacher's eye but not to Coker's dignified demeanor, his slim figure, or his German-trained rigorous insistence that the student apply himself.

Coker encouraged students by simple acts of kindness. A young woman who had just defended her thesis for the master's degree recalled a thoughtful gesture by Dr. Coker, who at this time had retired but went daily to his office as a research professor. "When I had completed the oral portion of my final Master's examination, I was invited to 'wait in the hall.' Dr. Coker's office was just across the corridor. He invited me to join him in his office, chatted amiably and offered me delicious peanut brittle."

The same student recalled an impromptu field trip that involved an invitation to "The Rocks." "On Dr. Coker's front lawn, there was a splendid, complete 'fairy ring' of mushrooms. He arranged to take us in his car from the campus to his home to see the beautiful display—at 4:00 A.M.! This time was appropriate because if we waited until sunrise the mushrooms would begin to change, because of the effect of the sunlight, from their pristine white to an undistinguished tan-gray. The capstone of the morning was the delicious breakfast prepared by him and Mrs. Coker, with the pièce de résistance being a delicious mushroom soup!"[24] In fact, Coker was greatly interested in the field trips of his students. One student recounts that every Monday morning Professor Coker called him into his office to inquire of him a report on his collecting trip of the weekend.[25]

In the days before financial aid, when hard times threatened the progress of a student's education, Coker quietly made it possible for individual students to remain at the University both by supporting the University student loan fund established during the Depression and by responding directly to individual appeals.[26] His letter to a relative in Fayetteville, North Carolina, illustrates Coker's concern for the education of talented young people. He offered lengthy advice for his cousin's literary son, whom she feared was trapped as a hardware store clerk. "Robert," Coker wrote to the boy's mother, "should do everything in his power to get an education." He told her that the University made available to students certain long-term loan funds. A large number of students could make expenses while working. Board was available for students waiting on tables in the commons hall. Robert should write the University at once and prepare himself to enter the following year. Coker offered to give him a financial boost: "Should he get through High School and be inclined to help himself as much as possible," Coker wrote, "I shall be glad to lend him $100 a year."[27]

Another offer to a student was more closely related to Coker's own situation. In 1917, J. I. Somers of Burlington asked the professor about a job that would enable him to continue his studies. Professor Coker replied to his inquiry by offering him a room in his house to share with another student. He also suggested that he pay for his room at the rate of $5 a month by working on his place. Coker would also give him other work on his farm to bring in extra money, work that could involve milking one or two cows. With four hours a day of work, he could make $12 a month in addition to room rent. "Though this is not much," Coker wrote, "it is all I have at present and may get you through if you have some resources of your own to start with."[28] Toward the end of his career, in the 1940s, Coker made it possible by a loan for a student in his department to make the trip to Nova Scotia to observe *in situ* the large pitcher plant (*Sarra-*

This portrait of William Chambers Coker (which hangs in William C. Coker Hall at Chapel Hill) was painted by Alice Kent Stoddard and was a gift of Mrs. Louise Venable Coker. *Permission of Campus Historical Committee and courtesy of the photographic laboratory in Wilson Library.*

*cenia*) that he was studying.[29] By these and other ingenious suggestions and actions, Coker gave his students the push they needed to finish at the University or otherwise to advance their studies.

Letters from former students reveal that Coker's encouragement was more personal than simply the offer of loans and jobs. Rex W. Perry wrote him from Toronto in January of 1918, "I not only want to see the 'Hill' again but more especially yourself and to tell you how grateful I always have been and will be to you for your assistance to me at a time when it meant so much."[30] E. Oscar Randolph, who was then teaching in the Department of Geology and Biology at Elon College, included this comment in a "bread and butter" letter written after a visit in the home of his old professor: "Your continued openness and freeness, yet at the same time firmness and dignity, cause me to regret more and more that I did not enjoy more of your instruction and fellowship while a student in the University. Your keen interest in me, my home, and in my professional life will be highly treasured daily."[31]

After his retirement in 1944, Coker maintained an active interest in former students, many of whom considered him a friend. Returning from World War II in the mid-1940s, a student who had earned his master's degree in botany told of visiting his old professor to discuss his future. He expected to remain at Chapel Hill for his doctorate in botany, but Coker encouraged him to seek broader experience. This recollection demonstrates that the student's growth was more important to his former professor than the prestige of the academic department he had founded, where he always hoped to attract the brightest and best. This remembrance is worth quoting at some length.

When I came back from the wars, he of course had retired officially as head of the department which he had founded so many years before and John Couch was now head of the department. I still visited widely with Dr. Coker, however, and he was my good friend. In discussing my future, he said he felt that it was important to have a wide range of experience. I think he used the word "cross-pollination." It would never have crossed my mind to go to Harvard. I was very happy to stay in the lotus land of Chapel Hill, but he urged and communicated and pulled strings until I was awarded a teaching fellowship in the Biological Laboratories at Harvard, and until Irving Bailey, distinguished plant anatomist, agreed to accept me as his last graduate student. It was with great reluctance that I began to consider my departure from Chapel Hill, and for awhile it looked as if I would not be able to go. I kept getting a notice from the registration office, or something of the sort, at Harvard saying that my record was incomplete. I would write back and ask in what way it was incomplete, simply to get another statement that it was incomplete.

Finally, one day Dr. Coker said, "This is nonsense." He telephoned the registrar's office at Harvard to discover that that institution was apparently unique in requiring that the transcript should be sent in a typed copy which had been notarized. Otherwise, it was not considered valid. Had he not made the telephone call, I would not have been able to achieve my fellowship and go to Harvard to study. That, you might say, was one of his very last deeds of munificence to me.[32]

In several cases W. C. Coker was able to help his students find employment and, at the same time, supply candidates for positions in family-related enterprises in Hartsville. Older family members recalled that it was Will who suggested to his brother David the name of a star student of the Class of 1913 for the position of secretary-treasurer of Coker's Pedigreed Seed Company and J. L. Coker and Company in Hartsville. This student was A. L. M. Wiggins of Durham. Though his entire working career was in Hartsville, Wiggins later became president of the American Bankers Association, undersecretary of the Treasury for President Truman, and chairman of the Atlantic Coastline Railway.[33]

Among other persons whom Professor Coker recommended for work in Hartsville were W. L. Goldston, who became assistant manager of the Carolina Fiber Company in Hartsville during the summer and fall of 1916,[34] and Curtis Vogler, a graduate in botany at UNC who became in 1917 a plant breeder for Coker's brother David's Pedigreed Seed Company. Other UNC students whom W. C. Coker recommended for employment in Hartsville include master's students in biology Robert McIntyre and Waring Webb, and his Ph.D. students Velma Matthews, Budd Smith, and Arthur William Ziegler, all of whom became biology teachers at Coker College. Dr. C. Ritchie Bell, mentioned above, worked in Hartsville with Mrs. David R. Coker at Kalmia Gardens and later also at the Pedigreed Seed Company. The concerns of W. C. Coker for the employment of his students and also for the needs of his family in Hartsville worked at times to their mutual benefit.

Family support was not one-sided. David Coker, at the request of his brother in Chapel Hill, temporarily employed Dr. T. G. Harbison to find and locate unusual native plants for the newly established Kalmia Gardens in Hartsville in 1934. This was during the period when the University could not immediately find funds for Harbison's salary, even though he had already been officially engaged to organize Ashe's herbarium, which the University had miraculously acquired in late 1932.[35]

Coker was especially concerned with those of his students who served in the armed forces in 1917 and 1918. He corresponded with several of them. Though in age he was approaching the mid-forties, he himself took part in military training at UNC in the fall of 1917. He tried to join the war effort as a Red Cross rehabilitation worker in France. However, after many months of waiting, he received his classification too late to serve,

William Chambers Coker
with his namesake, William
Chambers Coker II, taken
in 1942. *Courtesy of the Coker
family.*

just two weeks before the armistice.[36] Later, Coker corresponded with his
students in wartime service during the World War II years of 1942–45.[37]

### Totten and Couch, Botanical Ambassadors in France

Both Henry Roland Totten and John Nathaniel Couch, enthusiastic students of botany at Chapel Hill, served in 1918 in the U. S. Army's Expedi-
tionary Force in France. Totten wrote Coker on July 11, 1918, just before
leaving for France, "I will always remember your kindness to me. You
have been more of a teacher than you know."

After the war, both Totten and Couch returned to Chapel Hill to become assistants in the botany department and eventually to receive the
Ph.D. under Dr. Coker. In 1916, Totten had already helped design and
develop the nationally recognized drug garden in the Arboretum. Upon
his return from France in 1919, where he studied medicinal botany, he became the garden's curator. He regularly used the Arboretum's drug garden as a teaching resource for students of pharmacy.[38] Totten was coauthor of Coker's tree books, the first of which was published in 1916.[39]
The Totten Center at the North Carolina Botanical Garden, built with a
bequest of the Tottens and dedicated in 1976, still houses at this writing

The Drug Garden in the Arboretum, August 5, 1923. *Permission of the North Carolina Botanical Garden.*

the Garden's offices, work areas, exhibition hall, and Totten Botanical Library. A living collection of medicinal plants in the Botanical Garden commemorates Totten's contribution to the University.

The John N. Couch Biology Library in Coker Hall is named for and dedicated in honor of one of the University's most distinguished graduates. Both Totten and Couch spent their professional lives teaching at Chapel Hill.

As a means of solving the serious logistical problem of mass troop transportation back to the United States after the armistice of November 1918, the U.S. government gave American officers the opportunity to study in France for some months. Totten and Couch accepted the offer with alacrity. John Couch studied botany and French at the University of Nancy, while Roland Totten worked in laboratories in Paris and studied botany and chemistry at the École de Pharmacie and the Institut Pasteur.

The letters exchanged between these two students and their professor during this period in 1919 took on a purposeful tone, even one of urgency, as Coker encouraged them to meet the best-known French mycologists, to visit their herbaria, to obtain specimens from them, and to collect in the field with them if at all possible. Both of Coker's students became botanical ambassadors without portfolio in the cause of mycology. In a letter dated March 24, 1919, Coker assigned Totten to collect as complete a series of French *Clavaria* as possible. He urged him to meet Narcisse Patouillard of Paris and René Maire of Caen, "the two most skilled living mycologists, both of whom have described species of *Clavarias*." He instructed Totten to obtain from them authentic specimens and to collect materials, submitting them while fresh to these experts for identification. "Try to get on a field trip with either or both of these botanists," he urged. Coker then listed twenty-four species of *Clavaria* he especially needed. He urged him to obtain species of *Clavaria* from Bresadola, the world-renowned Italian mycologist.[40] Coker also was eager to collect for the University at Chapel Hill certain rare mycological books. This list of books, which he hoped Totten could purchase in Paris for the University library, included a dozen rare works on fungi. He asked Totten to draw on him personally, to the extent of $1,000, to obtain these books.

Had he not been a hardy and energetic individual, Totten might have been overwhelmed by these assignments, even tempted to throw up his hands. However, he plunged into the work with good grace. He wrote Coker on April 22, 1919, that he would try to accomplish at least a part of his commission.[41] He described the formidable schedule for his studies: he worked nearly half-time in the laboratory of Dr. Matrouchot while continuing laboratory work on medicinal plants at the pharmacy school and also taking the lecture course in chemical biology at the Pasteur Institute. In addition, he spent three mornings a week in Professor Guenin's laboratory, where he had access to an excellent medicinal garden and to greenhouses and where he accumulated a good collection of prepared slides. He reported that Professor Matrouchot would assist in the book search, though he thought it impossible to find some of the rarer books. Totten informed Coker that Bresadola was still living in that part of Italy (Trento) recently retaken from Austria. Maire was in Algiers. He doubted being able to see either of them.

Coker wrote Couch on April 2, 1919, that he hoped he was making a good collection of *Saprolegniaceae*. He especially needed *Saprolegnia myxta*, "one of the most uncertain of our species." He told Couch of his asking Totten to bring back species of *Clavaria*. Enclosing the list of rare books he had sent to Totten, Coker asked Couch to participate in the book search.

In a letter dated April 27, 1919, Totten described to Coker a most pleasant botanical trip to Fontainebleau, location of the Sorbonne's field laboratory. Couch, in a letter to Coker dated May 2, 1919, thanked him for his papers on *Amanita* and *Saprolegniaceae*. He reported having given the list of books that Coker wanted to the president of the Botanical Society, who had most of the books in his own library. All could be obtained, he thought, except the Saccardo, *Sylloge Fungorum*, a very rare work. Couch told Coker that he had written Totten about the books and had given him the address of a bookstore in Paris.

Coker's letter to Totten on May 7, 1919, listed bookstores in France, Germany, England, and Italy. He asked for a progress report. Writing one week later, Coker assured Totten of employment if he returned by mid-September. He authorized him to spend several hundred dollars to acquire authentically named specimens of any French *Agarics* or *Hydnums* or higher fungi like puffballs. This sum he would provide in addition to the $1,000 for books.

A postcard from Couch in Dijon, written May 21, 1919, told Coker of his plans to visit as many botanical gardens as possible and to see Totten in Paris about the book search. Couch wrote again on June 11 that he had met Sartory, one of the foremost French mycologists, who had looked over Couch's own collection of *Saprolegniaceae* and had identified one strain as *Saprolegnia mixta* and another as *S. ferax*.

Totten, in his last letter to Coker from France, written on June 26, 1919, listed the books included in his last package, books by Patouillard, by Sartory and Maire, by Sartory alone, by Fries, and by Barla. He reported that the Gillet could be had in six months.

Coker saw to it that not a day of his soldier students' study leave in France was wasted. These two enthusiastic students of Coker's were thus of great help to him in the completion of his two major works in progress, *The Saprolegniaceae* and *The Clavarias of the United States and Canada*, to be published in 1923 by the newly established University of North Carolina Press.

Totten and Couch, while meeting distinguished mycologists and pharmaceutical botanists in France on the recommendation of their professor, were simultaneously taking giant steps in their own education. They became Coker's representatives, thereby gaining self-confidence and motivation for the rigorous doctoral studies they were soon to undertake. Furthermore, they were making important additions to the botanical library and the Herbarium for their University. Coker's belief in the capability of these two students and his own enthusiasm for their mutual efforts energized them to become significant contributors to the science of botany.

Yet Coker did not confine his interest only to budding professional botanists. As a true teacher, he poured his energies into the development and welfare of many young people who for forty years passed through his classroom and laboratory in Davie Hall at UNC and who accompanied him on field trips. In addition to the scientists whom he shaped, his legacy also includes countless other students who caught his infectious love for the world of plants and for the beauty and wonder of nature. Many of these, who went on to other careers, remained grateful to have studied with him.

# CHAPTER NINE

# Writer and Editor

> *You have succeeded in blending the technical and the*
> *popular in an ideal way and, so far as I know, it is unique.*
>
> —J. C. Holmes to W. C. Coker

WILLIAM CHAMBERS COKER arrived at Chapel Hill eager to
continue research and writing. Within five years after his appointment to the faculty of the University of North Carolina, he
had published seventeen articles plus his dissertation, the first publication of the Botanical Laboratory of Johns Hopkins University.[1] The *Botanical Gazette* of the University of Chicago published six of his articles; the
*Journal of Applied Microscopy and Laboratory Methods,* two; the *Journal of
the Elisha Mitchell Scientific Society,* seven (including abstracts); and *Torreya,* two. In 1905, a book on the Bahama Islands included Coker's long
article on the islands' vegetation.[2]

He early set about recording the flora of his new surroundings. In 1903,
during his first academic year at the University, he published "The Woody
Plants of Chapel Hill;"[3] in 1904 "Chapel Hill Liverworts;"[4] and in 1907,
"Chapel Hill Ferns and their Allies."[5] Within the next decade, he had pub-
lished with H. R. Totten "The Shrubs and Vines of Chapel Hill"[6] and *The
Trees of North Carolina.*[7] The list of Coker's publications that appears in
Appendix 1 here reveals both the number and variety of journals that pub-
lished his work and the breadth of his interest in botanical subjects.

Early in his career, Coker occasionally used biblical imagery for emphasis. In his presidential address before the North Carolina Academy of
Science in 1910, he advised against teaching subjects in isolation, subjects
that in his opinion were not relevant to the lives of students and that

W. C. Coker in front of
*Sequoia washingtoniana*,
Yosemite, 1909. *Southern
Historical Collection, Wilson
Library, University of North
Carolina at Chapel Hill.*

could distract them from useful pursuits. He wrote, "I do not mean to be irreverent when I call your attention to One who separated himself from the world and withdrew into the wilderness." Then he adds, "It is not everyone who would have the strength to say, 'Get thee hence, Satan.'"[8] In a 1909 article detailing a visit to Yosemite, Coker pays a lyrical tribute to the wonder of the sequoias: "As with all great objects and scenes, the mind cannot conceive for a time the full report of the senses. Indeed, the imagination can never fully grasp the fact that this towering mass before us was in its prime in those forgotten days when the psalmist asked, 'What is man, that thou art mindful of him? And the son of man, that thou visitest him?' Gradually the impression deepens until there falls upon us a sense of awe that cannot be produced by any of the temples or monuments that are built with hands."[9]

Often he used an informal colloquial style of writing. Early in his career, he expressed his affection for the plants of the sandhills near his home in South Carolina. He wrote of the "pretty little dwarf flowering locust (*Robinia nana*)" and also of the "pretty little partridge berry (*Mitchella repens*)." He considered smilax "one of the most beautiful evergreen vines in the world."[10] In his article on finding the grave of the botanist Thomas Walter, he pays tribute to *Smilax walteri*: "What more charming memorial could one desire than the lovely wreaths of this cardinal of the woods that brighten the cold swamps with such a glowing flame?" He then added a personal note: "Every Christmas, at our home in Hartsville, we go out into the swamp and bring in these brilliant berries to add color to the day."[11]

In a letter to Coker dated June 25, 1934, his friend, State Forester J. C. Holmes, commented on his writing style in the recently published *Trees of the Southeastern States*: "You have succeeded in blending the technical and the popular in an ideal way and, so far as I know, it is unique." Coker's description in the 1945 edition of this book of a stand of Canadian hemlock (*Tsuga canadensis*) on a north-facing bluff of Swift Creek in Wake County, North Carolina, illustrates this style. After remarking that this mountain conifer is seldom found in the central part of North Carolina, Coker described the location and size of some of these hemlocks. He then added his own ardent hope for them: "May they be spared another fifty years."* Happily, his wish has been more than granted. The hemlocks are now thriving nearly seventy years after Coker's prayer for their survival.

His description of the big-fruited haws (*Crataegus ravenelii*) of "Haw Ridge" in Chesterfield County, South Carolina, is another example of Coker's folksy style. He calls these trees "the most remarkable collection of hawthorns in a limited area that we have ever seen. . . . Within a few feet of Mr. Sowell's garden [on his farm near McBee, South Carolina] occurs one of the largest hawthorns in diameter on record and very near to it is another specimen almost as large that bears the largest fruit of any known species in the area covered by this book. There are within a few hundred yards on the same ridge six or seven other trees of the same species bearing equally large fruit. It is these big-fruited haws that make Haw Ridge famous in this section of South Carolina, as they are prized for eating and the making of jelly." He added that the largest of these trees of *C. ravenelii* have a fruit that "though considerably smaller is still good eating." He then reported the remarkable fact that the seeds of these haws germinate the first spring after planting and that, moreover, the haw's

---

* *Trees of the Southeastern States*, 3rd ed. (Chapel Hill: U of North Carolina P, 1945) 37. The older trees are doing well and there are abundant smaller trees at this site, which is now preserved as "Hemlock Bluffs Nature Preserve" in the Stevens Nature Center of the Cary Department of Parks, Recreation and Cultural Resources. These trees were registered with the Capital Trees Program in the "Landmark Tree" category in 1996. Laura White, Stevens Nature Center supervisor, provided this information in May of 2000.

T. G. Harbison and May R. Coker beside a giant *Crataegus ravenelii* at Haw
Ridge, Chesterfield County, South Carolina, ca. 1934. *Permission of the Rare
Book Collection, University of North Carolina at Chapel Hill.*

characteristics "are usually perpetuated in its offspring." Mrs. Sowell's sister, Coker recounted, took seed to her home in Georgia; these immediately sprouted and "are now bearing fruit just like those of the home tree."

Coker's description here is far different from the usual botanical writing. The reader has the impression of accompanying the curious botanist
on a field trip to Haw Ridge in the sandhills near Hartsville, of helping
him measure the trunks, of tasting the fruit, all while absorbing both
what Coker sees and what Mrs. Sowell is telling him about how quickly
the seeds sprout and how successful her jelly is. Coker's records of observations in his tree book, though careful and factual, often have the flavor
of the southern folk narrative, a familiar trait in the rural culture of the

Carolinas in which he was raised. As a writer he successfully mixed the language of science and home.

Coker published three obituaries of botanists, all friends whom he admired—namely W. W. Ashe, Lars Romell, and Duncan Starr Johnson. The home background of W. W. Ashe was comparable to his own. Ashe was four months older than Coker. He was the son of a Confederate officer, as was Coker, and a member of a large family in a sizable home surrounded by gardens and woodlands. Coker wrote of Ashe's home, "Here were many kinds of native and cultivated trees, shrubs, and fruits, an ideal refuge for a multitude of birds." Ashe's surroundings throughout his youth, like Will Coker's, invited observation of plant life and other natural phenomena. Coker and his coauthors of the Ashe obituary, J. S. Holmes and C. F. Korstian, paid tribute to Ashe as "an indefatigable observer, collector, and annotator of plants. Few, if any, were his equal in first-hand knowledge of the flora and vegetative types of the southeastern states, more especially in the woody plants and of the less accessible areas." To Coker, Ashe was "a man of transparent honesty, unselfish devotion to duty, happy and cheerful in his work and always appreciative of the work of others." [12] In his tribute to his friend, the Swedish mycologist Lars Romell, Coker remarks on his unusually powerful and accurate mind, his elevation of character and beauty of spirit that deeply impressed all who knew him.[13] In paying tribute to Duncan Starr Johnson, his professor, dissertation director, and close friend[14] from his years at Johns Hopkins, Coker seems to reflect upon Johnson's influence on his own life: "The great breadth of Johnson's botanical interest is illustrated by his excursions into the field of ecology and plant distribution. He never expressed a mean or ungenerous thought and I do not believe that he ever entertained one. Throughout four years of close association with him as student and friend, the writer never knew him to show any sign of anger or undue excitement. His life seemed to run smoothly always, like a gently flowing river, too deep to be agitated, ever becoming deeper and broader as it ran its appointed course." [15] The life of unruffled peace of his old professor who later became his colleague was an ideal to which Coker aspired. Indeed, it is one he seems in large measure to have attained. In honoring the character traits of these three friends, Coker holds up a mirror to himself.

### Coker as Editor

William Chambers Coker was editor from 1904 until 1945 of the *Journal of the Elisha Mitchell Scientific Society*, founded in 1883 at Chapel Hill. One of his students, who later became a professor of botany at Chapel Hill, re-marked that Dr. Coker *was* the *Elisha Mitchell Journal*.[16] Coker's work as editor demanded considerable time and effort. There is preserved a con-stant exchange of letters between Coker and the printers of the journal, located first in Durham and then in Raleigh. He frequently had to urge his publisher to bring out the current number more promptly. He en-couraged potential authors to submit articles for the journal. He cajoled contributors to send an article or to return corrected proof in a timely fashion. Subscribers barraged him with complaints that their current number had not arrived or was late. He himself published many articles in the journal, among which are his richly illustrated book-length article of 1917 on the *Amanitas*,* a lengthy paper on the lower *Basidiomycetes* in 1920, and a fifty-page article on the *Thelphoraceae* in 1921.

The preparation of over fifty letters selected from the correspondence of the Kentucky physician-botanist Dr. Charles Wilkins Short (1794–1863) is perhaps Coker's most interesting editorial contribution. Un-doubtedly he relished the task, for these letters, which the family of Short donated to the University's Southern Historical Collection, are of prime interest to botanists. In this collection Dr. Short's most frequent corre-spondent was one of Coker's heroes, Asa Gray himself. Other correspon-dents represented in this collection of letters were such stellar botanists as John Torrey, Thomas Nuttall, and the Hillsborough, North Carolina, clergyman-botanist, Moses A. Curtis. The final letter of Asa Gray in this collection, to a daughter of Dr. Short, recounts the story of the search for the "Oconee bell," the lost *Shortia*, which Gray first encountered as a fra-gile dried specimen in 1839 while he was studying Michaux's herbarium in the Jardin des Plantes in Paris and which he immediately named for Charles Wilkins Short.[17]

W. C. Coker edited a volume of scientific papers by members of the University faculty entitled *Studies in Science*. This book was one of the University's sesquicentennial publications of 1949, a series directed by Louis Round Wilson. His work on this volume, his edition of the letters of the botanist Short, and his editorship of the *Elisha Mitchell Journal* were among Coker's many valuable contributions to the University of North Carolina.

Two excerpts from the 1945 edition of his tree book with Totten, one

---

* Professor Gudger, biologist at State Normal College at Greensboro, now UNC–Greens-boro, commented on Coker's *Amanita* article in a letter of October 8, 1917: "Your paper is a lasting monument to your patience, your industry and your wide knowledge of fungi. Our State is to be congratulated on the publication of such works by one of its citizens." SHC.

from the holly family and one from the tea family, illustrate Coker's deliberate departure from usual formal scientific writing style. In his discussion of the Dahoon holly, Coker permits himself to introduce an unpublished letter of Dr. J. H. Mellichamp to H. W. Ravenel, both nineteenth-century South Carolina botanists. Mellichamp is arguing for separating Dahoon holly (*Ilex cassine*) from myrtle-leaved holly (*Ilex myrtifolia*). The latter had up to then been judged to be a variety of Dahoon holly, but one that the writer had seen only in widely diverse habitats. Coker quotes Mellichamp as saying, "The former is found here, usually on the edge of wet springy 'salts'—the latter in shallow-clay pine barrens."[18]

Botanical mysteries, such as Gray's long search for *Shortia*, appealed to Coker and found their way into his writing. In his discussion of the tea family, Coker and Totten recount the story of the lost Franklin tree, *Gordonia alatamaha*,* which John Bartram and his son William had discovered in 1765 while botanizing on the Altamaha River in Georgia. The Native Americans called this river "alatamaha," the musical word retained in the botanical name. The Bartrams named this tree *Franklinia* in honor of their friend, Benjamin Franklin. In 1773, William Bartram returned and gathered seeds. In two subsequent visits, the younger Bartram again collected plants and seeds, which he introduced into cultivation in the Bartrams' Philadelphia garden and in England. Dr. Moses Marshall found this plant again in the same place in 1790. Since that time it has never been seen in the wild state, but it remains in cultivation in America and Europe.

Coker's choice of subjects and his unique writing style reflect his affection for the world of plants, his keen interest in the subject at hand, his tendency toward lyrical expression, his wry humor, and his willingness to depart from strict scientific description in order to interest the amateur botanist. His acceptance for so many years of the sometimes vexing editorial chores for the *Journal of the Elisha Mitchell Scientific Society* attests to his passion for encouraging scientific research and attracting public awareness to its importance. Coker's prolific writings, touched with his unique style, are an important part of his lasting legacy.

---

* After description in Coker and Totten, *Trees*, 1945, 343. Bartram's garden in Philadelphia calls this famous plant *Franklinia alatamaha*. See Garden Membership Bulletin, 1999. The botanical name uses the native spelling of the river, Alatamaha, not the current spelling, Altamaha. Coker would have been gratified to know that, as a celebration of the three-hundredth anniversary of John Bartram's birth in 1999, the Bartram garden in Philadelphia attempted to register all examples of the *Gordonia alatamaha* currently protected in American gardens. Thus, in Bartram's own garden, a distinguished plant was saved from near extinction and is now in wide cultivation.

# The Legacy of William Chambers Coker

WILLIAM CHAMBERS COKER'S passion for botany was his vital force. Nothing botanical was beyond his concern. Woven into the lives of individuals and into the fabric of institutions, his influence is recognizable fifty years after his death. Mycologists consider him among the distinguished contributors to this field. His studies of water molds and fleshy fungi are classics. Scholars and amateur botanists still refer to his publications on trees and on vascular and herbaceous plants. Students, faculty, staff, and visitors enjoy today the peace and beauty of the Coker Arboretum on the UNC campus and learn from its botanical diversity. The University takes pride in the Herbarium at Chapel Hill which is among the best in the United States. Scholars here and worldwide regularly use this priceless collection that Coker built steadily during his tenure at the University and that his successors have enriched and tended. He was among the founders of the Highlands Biological Station, which attracts graduate students and advanced scholars for the study of the plants and animals of the Appalachian Blue Ridge. The North Carolina Botanical Garden is currently fulfilling Coker's dream of heightening public awareness of the rich diversity of native plants in the state's varied ecological regions and in its local forests for the education and enjoyment of visitors to Chapel Hill. The role of the North Carolina Botanical Garden in education for the protection of natural land, a role Coker modeled, is of capital importance at this time when our natural woodlands vanish at an accelerating pace.

Coker had a remarkable ability to turn ideas into action. Repeatedly grasping and pursuing potentially useful ideas, he was able to transform them into something larger. Several projects may serve as examples. (1) A casual remark of President Venable to Coker during a campus stroll to their offices at the University in 1903 set the new botanist to work draining a five-acre boggy meadow and planting the drier northern side. These acres became the campus garden now known as the Coker Arboretum, which celebrates its centennial in 2003. (2) His father's request in 1910 that Will write a paper about the climate and plants of Hartsville for presentation at a meeting of the Pee Dee Historical Society led Will Coker to make a thorough study by ecological communities of local plants. This study resulted in the publication in 1912 of his first book, *The Plant Life of Hartsville*. (3) An invitation by the secretary of the North Carolina Academy of Science to speak at the annual meeting of the society on the subject of science teaching in the high schools prompted Coker to write and deliver in 1910 his presidential address on his philosophy of education, a talk containing ideas still useful for educators today. (4) Coker's early concern for the improvement of public school grounds motivated him to offer lectures in 1910 for meetings of the Woman's Association for the Betterment of Public School Houses and later to volunteer as a University extension agent for landscaping North Carolina public school grounds. He responded to requests from at least twenty North Carolina schools and published two extension bulletins with illustrations for planting individual school grounds. (5) Coker respected the professional ability of his personal friend John Nolen, the distinguished city planner from Cambridge, Massachusetts. He was convinced that Nolen should design a master plan for his University's campus. After the University selected Nolen, Coker worked closely with him and later helped to implement a design that incorporated many of Nolen's ideas. (6) Coker had firsthand knowledge of the extraordinarily rich plant life of the Highlands area of North Carolina's Blue Ridge. It took only a suggestion by Clark Foreman, president of the Highlands Museum, to prompt Coker to apply in 1928 for a desk in the museum in order to pursue his study of fungi and trees of the area. From this small beginning and from his annual summer work there grew his enthusiasm, along with that of other biologists and citizens of the town, for the development of the Biological Station at Highlands, North Carolina.

Coker constantly observed the natural world, his preferred milieu and his ever-present inspiration. His senses were fine-tuned to observe his natural surroundings. He steadily pursued the botanical research that his observations suggested to him, and he recorded his findings in his many publications. He maintained a lifelong interest in landscaping with attractive plants adapted to specific sites, especially in the Carolinas, where he well knew the varying climatic and soil conditions. His writings continue to guide those interested in the plant life of North Carolina and of the South. Perhaps the primary motivation for W. C. Coker's many activities was his reverence for the natural world and his belief in the "civilizing influence" of the individual's contact with nature. Coker wrote in 1921, after quoting Goethe, Byron, Chesterton, Hawthorne, and Tagore, "It is not possible to overestimate the ennobling influence of things that are beautiful and pure. They can strengthen and sustain beyond all power save human love. Encompassed and uplifted by the glory of the world, Whitman exclaimed: 'I am larger, better than I thought; I did not know I contained so much goodness.' This expansion of spirit before the pageantry of nature was proof of his own greatness, for 'The perception of beauty is a moral test.'"[1]

This belief of Coker's in the beauty of the local environment as an ennobling force for the individual and for the community not only lives on but is also today increasingly expressed in the lives of those who knew him and his work. Conservation of natural areas, woodlands, wetlands, and habitats where unusual plants flourish in the wild has become a national movement. It is a fulfillment of Coker's hopes that preservation of nature is particularly strong in North and South Carolina, the states to which he contributed most. In spite of efforts at preservation, many vital areas have already been lost. Some of Coker's writings that draw attention to deposits of rare plants must now be valued as history. The bulldozer has peeled away areas where he collected and led field trips. Conservationists and concerned citizens, however, have protected certain locations mentioned in Coker's writings including the "Hemlock Bluffs" of Wake County, North Carolina, now in the town of Cary's Hemlock Bluffs Nature Preserve, and the stands of venus flytrap localized near the coast.[2] Many today share William Coker's enjoyment of nature study in the field. On the increase are birdwatchers, wildflower enthusiasts, walkers and hikers, and citizens determined to protect precious lands for those who follow. As the state grows more populous, its wild places attract ever more people in search of the quiet enjoyment of nature.

The maintenance of pleasant roadsides interested Coker. Conditions of early twentieth-century roadsides demanded an approach much different from that of designers of the massive planting of flowers and shrubs that today border our high-speed highways. The slower-moving cars, the light traffic, and the winding two-lane roads of his day permitted a driver the luxury of more than casual looks at roadside beauty. Coker advised keeping inviolate the attractive closely wooded borders of eastern Carolina roads as well as the long vistas of the piedmont. But his advice was practical about what could and could not be done to make roadsides pleasing for the traveler. He did not recommend indiscriminate planting of roadside trees that could not regularly be tended.*

Preservation of unspoiled natural areas was a dream of Coker's, indeed a passion of his during the 1930s when he wrote many letters to congressmen, administrators in Washington, and foundations in an attempt to save the "primeval forest" near Highlands. The failure of this effort was a heartbreak for him. The need he foresaw to protect natural areas is even more urgent today as population pressures increase. The private preservation movement, still in its infancy during Coker's lifetime, is much stronger today in the Carolinas thanks to the vigorous activity of the several local, state, and national conservation organizations. It is also today more urgently needed.

In 1927, Coker wrote to UNC president Chase, "It is proposed to establish at Chapel Hill on University land a collection of all trees and shrubs native to North Carolina and to select for experimentation their most promising variations and forms."[3] Seventeen years later, in an unpublished paper of 1944, he restated this proposal.[4] The North Carolina Botanical Garden, its first nature trail opening in 1966,[5] could be considered a child of this dream of William Chambers Coker. Its first director was C. Ritchie Bell, who studied botany at Chapel Hill as an undergraduate toward the end of Coker's teaching career and later received the doctorate from the University of California at Berkeley. The current director, Dr. Peter White, an authority on wildflowers, and his associates carry on Coker's ideal of conservation and propagation of our native plants and trees and education of the public about their heritage of the fields and forests.

Coker did not simply put native trees and shrubs into collections; he believed in using them in the landscape. The variety of shrubs and trees and the placement of plants around the newly constructed UNC buildings while he was chairman of grounds and buildings attest to his knowledge of where various plants would thrive and to his aesthetic taste in using them. The historic central campus of the University of North Carolina at Chapel Hill, which has characteristics of an arboretum in the diversity of

---

* Writing on November 28, 1928, in response to Miss Mary Hyman's inquiry about having children plant the roadsides of Orange County, Coker gave a number of practical suggestions. He advised the preserving of distant views in the piedmont and mountain sections of the state and letting natural woodland borders serve roads through forested areas. Roadside trees planted in agricultural lands would be pleasing if well tended, though this was seldom the case. With these thoughts in mind, he could not advise indiscriminate planting of roadsides by schoolchildren or others. SHC.

its trees, is a memorial to Coker's work in guiding the planting for the University. Others have continued to enhance and protect this diversity. Coker collected native shrubs in a well-designed garden on University property near the present Mason Farm Biological Reserve. This effort was interrupted, but it continues today at the North Carolina Botanical Garden.

Hartsville, South Carolina, which Coker always considered home, bears his mark. Though John Nolen visited Hartsville in 1915 and made suggestions for the Coker College campus by submitting to Major Coker and to Will what he termed a "rough sketch," it was the younger Coker who planned the landscaping there and oversaw the planting. Several of W. C. Coker's former students became biology teachers at the college in Hartsville.[6] The William Chambers Coker Science Building at Coker College is in large measure the result of his generosity. W. C. Coker's strong desire to preserve the unusual natural flora on the banks of Black Creek west of Hartsville motivated him to buy acreage there in 1932, during the depth of the Depression, and to deed those acres to his sister-in-law, an enthusiastic gardener. This land, together with additional acres purchased by the new owner, is now known as Kalmia Gardens, the botanical garden of Coker College.

Coker became a world-recognized authority on fungi, but his interests could not be contained in only one area. Though his study of fungi was lifelong, he early studied seed development in conifers. He wrote articles about many herbaceous and woody plants. He documented the range and variety of native trees and shrubs, especially in the Carolinas. He readily offered to others, on request and without compensation, his considerable talent as an amateur landscape designer. Having no need for accolades as a specialist in one particular field, he pursued his diverse botanical and civic interests, which included landscaping projects for the public good.

Though professional and public recognition was of little concern to him,[7] Coker accepted several responsibilities in scientific organizations. A year after his arrival at the University, he was asked by the Geographical Society of Baltimore to head the botany staff on an expedition to the Bahamas.[8] Two years after assuming responsibilities at Chapel Hill, he became editor of the *Journal of the Elisha Mitchell Scientific Society*. Under his forty-year editorship, the journal became internationally recognized. He was twice president of the Elisha Mitchell Society, in 1907–8 and in 1919–20.[9] Seven years after his arrival at the University, he became president of the North Carolina Academy of Science. In 1927 he was elected chairman of the mycological section of the Botanical Society of America, and from 1939 to 1941 he served as the first chairman of the southeastern section of that society. In 1935, he presented an invitation paper on the

William Chambers Coker with Mrs. Coker and his niece Elizabeth
Boatwright Coker at a wedding reception of another niece in Hartsville,
1949. *Courtesy of the Coker family.*

aquatic fungi before the mycological section of the Sixth International
Plant Congress in Amsterdam. Though appointed vice president of the
mycological section of the Seventh International Plant Congress at Stockholm in 1950, he was obliged to decline because of ill health.[10]

William Coker's good taste is a part of his legacy. He chose to build in
1908, on the property he had bought two years earlier, a home in a modified form of the prairie style of architecture, which Frank Lloyd Wright

was introducing at the time. The design was simple; the house was beautifully crafted and suited to the site. Dr. and Mrs. W. Woodrow Burns Jr. have restored and preserved his home, "The Rocks," and today are cultivating a garden around it. Dr. and Mrs. James L. Peacock are also tending parts of his garden, including a row of nineteen hollies (*Ilex opaca*) that Coker brought as small plants from the woods as an experiment to determine their sex and to demonstrate the great variety exhibited in native plants.[11]

Coker designed the grounds surrounding his house not with perennial beds but rather with a variety of native and exotic trees and shrubs. His garden not only gave pleasure to friends, visitors, and passersby but also served as an example of garden design for the community with plants that would flourish in the local environment. During his lifetime, the garden at "The Rocks" was a haven of beauty, that one of his students termed a "magic garden."[12]

William Coker had a taste for the art of his time. He bought from an art gallery in New York a statuette by Daniel Chester French called "Spirit of Life," the study for a larger model for the Spencer Trask memorial at Saratoga Springs, New York. This sculpture, W. C. Coker's gift to the University, now stands in a lighted niche at the first landing of the great stairway opposite the main north entry of Wilson Library at the University of North Carolina. Dr. Coker's friend Julia Booker wrote to the artist to tell him how happy she was to have his sculpture at the University. French replied in a letter dated February 14, 1924, "Not only am I made happier by your applause, but it gratifies me that your rich bachelor likes it well enough to go so deep into his pocket for it. That is one of the tests!"[13]

This graceful figure of a young woman, holding high a blooming branch in her right hand and the overflowing bowl of life and learning in her left, symbolizes the passion of William Chambers Coker, its donor, for plants and for all learning. When Wilson Library was the principal resource for undergraduate and graduate research at the University, this little figure gave hope and courage to the student climbing the great stairway to the card catalog and the stacks for concentrated study. Since Wilson Library has become the repository of special collections, the small illumined figure continues to delight those who enter there to use rare books and manuscripts or other special collections, or who pause on the landing to enjoy it on their way to view the changing library displays in the exhibition gallery above. This sculpture is one of the continuing legacies of William Chambers Coker.

A student of botany during the summer of 1953 remarked that on the day of Dr. Coker's burial Dr. Totten kept his students hard at work as a

William Chambers Coker's last photograph. *Permission of Couch Biology Library at the University of North Carolina at Chapel Hill.*

special tribute to his mentor.[14] Laurie Radford commented on the service on that July day, "The graveside service . . . in the old Chapel Hill Cemetery, conducted with simplicity and dignity, was attended by a great number of family and friends. Thus ended the days of a very talented man, whose life was filled with useful study and a wide range of accomplishments. He was always searching for the truth, always loving his work and the study of plants. No one knows how many times he gave a young person a helping hand or contributed generously to a worthy cause. He was quite handsome, with the look and bearing of an aristocrat, yet sincere and approachable. I count myself fortunate indeed to have known him both as teacher and friend, and to have had the opportunity and privilege of working with him for so many wonderful years."[15] Coker's influence upon students in his department was lasting.

One may well ask how William Chambers Coker managed to envision such a number of projects, undertake them, and bring them successfully to reality and also how he even managed to institutionalize several of these projects so that they have remained his gifts for those who have followed. Perhaps one reason for this extraordinary accomplishment is that his driving passion for the study of plants consumed almost entirely the energy of his early and middle years, years when many of his scholarly contemporaries were concerned with the demands of their households. He married late in life, after accomplishing much. He had no children. Moreover, he was often able by his own means to supplement the cost of his fertile ideas. One can only ascribe some of his success to good fortune. His parents encouraged him. Knowing the value of his work, his siblings could be depended on to collect plants for him when requested to do so. He married a lifelong friend, whose conscientious care for him allowed him to entertain friends in their lovely home and garden. She assumed his domestic concerns, thus providing for him the freedom to continue his research well into his final years. The acres of the Arboretum were too wet with underlying springs to afford a stable building site for the burgeoning University campus. He served the University under administrations that were visionary and receptive to the vision of others. And, perhaps most important, he worked in a state with unusual botanical diversity, containing diverse ecosystems from seaside to high mountains. The native forests of North Carolina encompassed more species of trees than all of northern and central Europe.[16] Some of these rich communities remained pristine during his lifetime.

The celebration in 2003 of the centennial of the Coker Arboretum on the campus at Chapel Hill calls to remembrance the contributions of William Chambers Coker—his devotion to science and the preservation of our natural heritage, his reverence for natural beauty, his gifts to his University and his state, his influence on his students, his many landscaping projects, and his practical ideals for education and its extension beyond academia. His productive life was that of a modest, hardworking, imaginative, humorous, kind, and generous man whose good taste and creative ideas remain alive today.

# Publications of William Chambers Coker

Based on compilation by John N. Couch and Velma D. Matthews in *Mycologia*, vol. 46; no. 3, May–June 1954.

1902 Notes on the Gametophytes and Embryo of *Podocarpus*. Bot. Gaz. 33: 89–107, pls. 5–7.

1903 The Woody Plants of Chapel Hill, N.C. Journ. E. M. Sci. Soc. 19: 42–49.

1903 Selected Notes I. Leaf Variation in *Liriodendron tulipifera* Bot. Gaz 35: 135–138, figs. 1–6.

1903 Selected Notes II. Liverworts Bot. Gaz. 36: 225–230, figs. 1–5.

1903 On the Gametophytes and Embryo of *Taxodium*. Bot. Gaz. 36: 1–27, 114–140, pls. 1–11.

1903 Algae and Fungi for Class Work. Journ. Applied Micros. and Lab. Methods 6: 2411–2412.

1903 A New Method of Sprouting Pollen Grains. Journ. Applied Micros. and Lab. Methods 6: 2495–2496.

1904 Selected Notes III. *Equisetum arvense* L.; Multiseeded acorns; *Clavaria mucida* Pers. Bot. Gaz. 37: 60–63, figs. 1–17.

1904 On the Spores of Certain Coniferae. Bot. Gaz. 38: 206–213, figs. 1–24.

1904 Chapel Hill Liverworts. Jour. E. M. Sci. Soc. 20: 35–37.

1904 Angiosperms with Exposed Ovules. (Abst.) Journ. E. M. Sci. Soc. 20: 117.

1905 Observations on the Flora of the Isle of Palms, Charleston, S.C. Torreya 5: 135–145, figs. 1–4. Abst. in Journ. E. M. Sci. Soc. 20: 12. 1904.

1905 Vegetation of the Bahama Islands. In G. B. Shattuck (ed.), The Bahama Islands, pp. 185–270, pls. 1 and 33–47. MacMillan, Geographical Society of Baltimore, New York.

1906 The Embryo-sac of *Liriodendron*. (Abst.) Journ. E. M. Sci. Soc. 22: 61.

1906 Liverwort Types for Elementary Classes. (Abst.) Journ. E. M. Sci. Soc. 22: 61.

1906 The Endosperm of the Pontederiaceae. (Abst.) Journ. E. M. Sci. Soc. 22: 61.

1907 Fertilization and Embryogeny in *Cephalotaxus fortunei*. Bot. Gaz. 43: 1–10, pl. 1 and 5 text figs.

1907 The Development of the Seed in the Pontederiaceae. Bot. Gaz. 44: 293–301, pl. 23.

1907 Chapel Hill Ferns. (Abst.) Journ. E. M. Sci. Soc. 23: 50.

1907 Chapel Hill Ferns and Their Allies. Journ. E. M. Sci. Soc. 23: 134–136.

1908 (With J. D. Pemberton) A New Species of *Achlya*. Bot. Gaz. 45: 194–196, figs. 1–6. Abst. in Journ. E. M. Sci. Soc. 23: 48. 1907.

1908 The Recent Baltimore Meetings of Scientific Societies (Botany Section). Journ. E. M. Sci. Soc. 24: 155–158.

1909 A Visit to the Yosemite and the Big Trees. Journ. E. M. Sci. Soc. 25: 131–143.

1909 Vitality of Pine Seeds and the Delayed Opening of Cones. Amer. Naturalist 43: 677–681. Reprinted in Journ. E. M. Sci Soc. 26: 43–47, 1910, and abst. in 25: 47.

1909 Additions to the Flora of the Carolinas. Bull. Torrey Bot. Club 36: 635–638. Reprinted in Journ. E. M. Sci. Soc. 25: 168–171.

1909 A Double-Flowered *Sarracenia*. Plant World 12: 25: 253, 4 figs.

1909 Liverwort Types for Elementary Classes. Torreya 9: 233–236, figs 1–4. Abst. in Journ. E. M. Sci. Soc. 22: 61. 1906.

1909 *Lycopodium adpressum* forma *polyclavatum* from South Carolina. Fern Bull. 17: 83–85.

1909 Some Rare Abnormalities in Liverworts. The Bryologist 12: 104–105. 2 figs.

1909 *Leptolegnia* from North Carolina. Mycologia 1: 262–264, pl. 16.

1910 A New Host and Station for *Exoascus filicimus* (Rostr.) Sacc. Mycologia 2: 247.

1910 Another New *Achlya*. Bot. Gaz. 50: 381–383, 8 figs.

1910 A Visit to the Grave of Thomas Walter. Journ. E. M. Sci. Soc. 26: 31–42, pls. 13 and 14.

1910 Science Teaching. North Carolina High School Bulletin 1: 35–52.

1911 Report of the Committee of the North Carolina Academy of Science on Science Teaching in North Carolina. North Carolina High School Bulletin 2: 122–126.

1911 Dr. Joseph Hinson Mellichamp. Journ. E. M. Sci. Soc. 27: 37–64, pl. 6.

1911 The Garden of André Michaux. Journ. E. M. Sci. Soc. 27: pt. 2: 65–72, pl. 1–4.

1911 The Plant Life of Hartsville, S.C. Journ. E. M. Sci. Soc. 27: 169–205, pl. 1–15. Also published in much enlarged form by the Pee Dee Hist. Assoc. 1912.

1911 (With Louise Wilson) *Schizosaccharomyces octosporus*. Mycologia 3: 283–287, pl. 55 and 2 text figs. Abst. in Journ. E. M. Sci. Soc. 27: 83.

1911 Additions to the Flora of the Carolinas-II. Torreya 11: 9–11.

1911 Some Interesting Water Molds. (Abst.) Journ. E. M. Sci. Soc. 27: 83.

1912 The Seedlings of the Live Oak and White Oak. Journ. E. M. Sci. Soc. 28: 41–43, pls. 2 and 3.

1912 (With O. W. Hyman) *Thraustotheca clavata*. Mycologia 4: 87–90, pl. 63.

1912 *Achlya deBaryana* Humphrey and the Prolifera Group. Mycologia 4: 319–324, pl. 78.

1912 *Achlya glomerata* sp. nov. Mycologia 4: 325–326, pl. 79.

1914 Two New Species of Water Molds. Mycologia 6: 285–302, pls. 146–148.

1914 The Botanical Society of America. Journ. E. M. Sci. Soc. 29: 113–115.

1914    An *Achlya* of Hybrid (?) Origin. (Abst.) Journ. E. M. Sci. Soc. 30: 63. Also in Science 40: 386.

1914    Some Rare Plants and Singular Distributions in North Carolina. (Abst.) Journ. E. M. Sci. Soc. 30: 66–67. Also in Science 40: 387.

1915    Our Mountain Shrubs. Journ. E. M. Sci. Soc. 31: 91–112.

1915    (With E. O. Randolph) Observations on the Lawns of Chapel Hill. Journ. E. M. Sci. Soc. 31: 113–119.

1915    Winter Grasses of Chapel Hill. Journ. E. M. Sci. Soc. 31: 156–161.

1915    The Lawn Problem in the South. Journ. E. M. Sci. Soc. 31: 162–165. Abst. in 30: 67. 1914.

1915    The Teaching of Botany in High Schools. North Carolina High School Bull. 6: 77–79.

1916    The Laurel Oak or Darlington Oak (*Quercus laurifolia* Michx.). Journ. E. M. Sci. Soc. 32: 38–40, pls. 2–5.

1916    Some Interesting Mushrooms. (Abst.) Journ. E. M. Sci. Soc. 32: 46–47. Also in Science 44: 360.

1916    (With H. R. Totten) The Shrubs and Vines of Chapel Hill. Journ. E. M. Sci. Soc. 32: 66–81.

1916    Campus. Alumni Review 4: 154–155.

1916    (With H. R. Totten) The Trees of North Carolina. (106 pp.) Pub. by W. C. Coker, Chapel Hill, N.C.

1917    The Amanitas of the Eastern United States. Journ. E. M. Sci. Soc. 33: 1–88, pls. 1–69.

1917    *Saprolegnia anisospora* in America. (Abst.) Journ. E. M. Sci. Soc. 33: 95.

1918    The Lactarias of North Carolina. Journ. E. M. Sci. Soc. 34: 1–62, pls. 1–40.

1918    A Visit to Smith Island. Journ. E. M. Sci. Soc. 34: 150–153, pls. 10–16.

1919    The Hydnums of North Carolina. Journ. E. M. Sci. Soc. 34: 163–197, pls. 1–29.

1919    A New Species of *Amanita*. Journ. E. M. Sci. Soc. 34: 198–199, pls. 30 and 31.

1919    A Parasitic Blue-green Alga. (Abst.) Journ. E. M. Sci. Soc. 35: 9.

1919    *Craterellus, Cantharellus* and Related Genera in North Carolina; with a Key to the Genera of Gill Fungi. Journ. E. M. Sci. Soc. 35: 24–48, pls. 1–17.

1919    The Distribution of *Rhododendron catawbiense,* with Remarks on a New Form. Journ. E. M. Sci. Soc. 35: 76–82, pls. 19–22.

1920    Notes on the Lower Basidiomycetes of North Carolina. Journ. E. M. Sci. Soc. 36: 14.

1920    Genera of Lower Basidiomycetes Not Before Reported from North America, (Abst.) Journ. E. M. Sci. Soc. 36: 14.

1920    *Azalea atlantica* Ashe and Its Variety *luteo-alba* n. var. Journ. E. M. Sci. Soc. 36: 97–99, pls. 1 and 7.

1920    (With J. N. Couch) A New Species of *Achlya*. Journ. E. M. Sci. Soc. 36: 100–101.

1920    (With H. R. Totten) Laboratory Guide in General Botany. Chapel Hill, N.C. Revised editions in 1926 and 1931.

1921 Notes on the Thelephoraceae of North Carolina. Journ. E. M. Sci. Soc. 36: 146–196, pls. 14–35.

1921 (With Eleanor Hoffman) Design and Improvement of School Grounds, pls. 1–20. Bur. of Extension Bulletin, Special Ser. No. 1. Chapel Hill, N.C.

1921 Some Fungi New to North America or the South. (Abst.) Journ. E. M. Sci. Soc. 37: 13–14.

1921 (With H. C. Beardslee) The Collybias of North Carolina. Journ. E. M. Sci. Soc. 37: 83–107, pls. 1 and 4–23.

1922 (With F. A. Grant) A New Genus of Water Mold Related to *Blastocladia*. Journ. E. M. Sci. Soc. 37: 180–182, pl. 32.

1922 A Visit to Lapland and to Some Old Herbaria. (Abst.) Journ. E. M. Sci. Soc. 38: 24–25.

1922 (With H. C. Beardslee) The Laccarias and Clitocybes of North Carolina. Journ. E. M. Sci. Soc. 38: 98–126, pls. 1 and 7–33.

1923 The Saprolegniaceae. (201 pp., 63 pls.) The University of North Carolina Press, Chapel Hill.

1923 The Clavarias of the United States and Canada. (209 pp., 92 pls.) The University of North Carolina Press, Chapel Hill.

1923 (With J. N. Couch) The Gasteromycetes of North Carolina [Phalloids]. Journ. E.M. Sci. Soc. 38: 231–243, pls. 71–83.

1923 (With J. N. Couch) A New Species of *Thraustotheca*. Journ. E. M. Sci. Soc. 39: 112–115, pls. 8.

1924 (With Enid Matherly) How to Know and Use the Trees. Extension Bull. 3: No. 14. pls. 1–39. Chapel Hill, N.C.

1924 The Geasters of the United States and Canada. Journ. E. M. Sci. Soc. 39: 170–224, pls. 18–36.

1924 (With H. C. Beardslee) The Mycenas of North Carolina. Journ. E. M. Sci. Soc. 40: 49–91, pls. 6–30.

1924 (With J. N. Couch) Revision of the Genus *Thraustotheca*, with a Description of a New Species. Journ. E. M. Sci. Soc. 40: 197–202, pls. 38–40.

1926 Further Notes on Hydnums. Journ. E. M. Sci. Soc. 41: 270–287, pls. 51–65.

1926 (With H. H. Braxton) New Water Molds from the Soil. Journ. E. M. Sci. Soc. 42: 139–149, pls. 10–15.

1927 Other Water Molds from the Soil. Journ. E. M. Sci. Soc. 42: 207–226, pls. 27–36.

1927 (With P. M. Patterson) A New Species of *Pythium*. Journ. E. M. Sci. Soc. 42: 247–250, pl. 46.

1927 New or Noteworthy Basidiomycetes. Journ. E. M. Sci. Soc. 43: 129–145, pls. 12–22.

1927 Lars Romell. Journ. E. M. Sci. Soc. 43: 146–151, 1 pl.

1928 The Distribution of Venus's Fly Trap (*Dionaea Muscipula*). Journ. E. M. Sci. Soc. 43: 221–228, pl. 33.

1928 Notes on Basidiomycetes. Journ. E. M. Sci. Soc. 43: 233–242, frontispiece and pls. 36, 37, 47, 48.

1928 The Chapel Hill Species of the Genus *Psalliota*. Journ. E. M. Sci. Soc. 43: 243–256, frontispiece and pls. 38–46 and 48.

1928   Variations in Our Native Ornamental Flowering Trees. (Abst.) Journ. E. M. Sci. Soc. 44: 19.

1928   The Limits of Life. (Abst.) Journ. E. M. Sci. Soc. 44: 48.

1928   (With J. N. Couch) The Gasteromycetes of the Eastern United States and Canada. (201 pp., 123 pls.) The University of North Carolina Press, Chapel Hill.

1929   Notes on Fungi. Journ. E. M. Sci Soc. 45: 164–178, frontispiece and pls. 10–23.

1930   The Flora of North Carolina. (Abst.) Journ. E. M. Sci. Soc. 45: 182–183.

1930   The Bald Cypress, Journ. E. M. Sci. Soc. 46: 86–88, pl. 7.

1930   Notes on Fungi, with a Description of a New Species of *Ditiola*. Journ. E. M. Sci. Soc. 46: 117–120, pls. 8 and 9.

1931   What Nature Gave to Carolina. Nature Magazine 17: 294–297, 345–347, illus.

1932   (With J. S. Holmes and C. F. Korstian) William Willard Ashe. Journ. E. M. Sci. Soc. 48: 40–47, frontispiece.

1932   (With H. R. Totten) Notes on Extended Ranges of Plants in North Carolina. Journ. E. M. Sci. Soc. 48: 138–140.

1933   The Opportunities for Botanical Study at the Highlands Laboratory. (Abst.) Journ. E. M. Sci. Soc. 49: 35.

1934   (With H. R. Totten) The Trees of the Southeastern States. (399 pp., 232 drawings, 3 pls.) The University of North Carolina Press, Chapel Hill.

1934   The Gasteromycetes of Venezuela. In Carlos E. Chardon: Mycological Explorations of Venezuela, chap. 20. The University of Puerto Rico, [Rio Piedras].

1935   A Remarkable New Rhododendron. Journ. E. M. Sci. Soc. 51: 189–190, pls. 53, 54.

1935   Parasitic Flowering Plants of North Carolina. (Abst.) Journ. E. M. Sci. Soc. 51: 249.

1936   Inter-relationships of the Saprolegniales. (Abst.) Proc. 6th Intern. Bot. Cong. (Amsterdam) 1: 268–270.

1937   Professor Duncan Starr Johnson. Science 86: 510–512.

1937   Blastocladiales, Monoblepharidales and (with Velma Matthews) Saprolegniales. North Amer. Flora 2, pt. 1: 1–76 (bibliog., pp. 69–76, with J. H. Barnhart).

1937   (With J. R. Totten) The Trees of the Southeastern States, 2nd ed. 417 pp. The University of North Carolina Press, Chapel Hill.

1938   (With Mary S. Taylor) Filmy Ferns in the Carolinas. Science 88: 402.

1938   (With Jane Leitner) New Species of *Achlya* and *Apodachlya*. Journ. E. M. Sci. Soc. 54: 311–318, pls. 38, 39.

1938   A Filmy Fern from North Carolina. Journ. E. M. Sci. Soc. 54: 349–350, pl. 40, figs. 5 and 6, and pl. 41.

1939   (With Leland Shanor) A Remarkable Saprophytic Fungoid Alga. Journ E. M. Sci. Soc. 55: 152–165, pls. 22, 23.

1939   A New *Scleroderma* from Bermuda. Mycologia 31: 624–626, one text fig.

1939   New or Noteworthy Basidiomycetes. Journ. E. M. Sci. Soc. 55: 373–386, pls. 43–44.

1941 (Edited) Letters from the Collection of Dr. Charles Wilkins Short. Journ. E. M. Sci. Soc. 57: 98–168.

1942 Notes on Rare Hydnums. Journ. E. M. Sci. Soc. 55: 373–386, pls. 34–44.

1943 *Magnolia cordata* Michaux. Journ. E. M. Sci. Soc. 59: 81–88, pls. 17–20 and 6 text figs.

1943 (With Alma Holland Beers) The Boletaceae of North Carolina. (96 pp., 66 pls.) The University of North Carolina Press, Chapel Hill.

1944 The Woody Smilaxes of the United States. Journ. E. M. Sci. Soc. 60: 27–69, pls. 9–39.

1944 The Smaller Species of *Pleurotus* in North Carolina. Journ. E. M. Sci. Soc. 60: 71–95, pls. 40–52.

1945 (With H. R. Totten) The Trees of the Southeastern States, 3$^{rd}$ ed. (419 pp., illus.) The University of North Carolina Press, Chapel Hill.

1946 The United States Species of *Coltricia*. Journ. E. M. Sci. Soc. 60: 71–95, pls. 40–52.

1947 Further Notes on Clavarias, with Several New Species. Journ. E. M. Sci. Soc. 63: 43–67, pls. 1–14.

1947 North Carolina Species of *Volvaria*. Journ. E. M. Sci. Soc. 63: 220–230, pls. 28–32.

1948 Notes on Some Higher Fungi. Journ. E. M. Sci. Soc. 64: 135–146, pls. 16–25.

1948 Notes on Carolina Fungi. Journ. E. M. Sci. Soc. 64: 287–303, pls. 37–54.

1949 *Colus Schellenbergiae* Again. Mycologia 41: 280–282.

1951 (With Alma Holland Beers) The Stipitate Hydnums of the Eastern United States. University of North Carolina Press, Chapel Hill.


## Reprints

1974 *The Boleti of North Carolina.* New York: Dover Publications (Reprint of the 1943 publication, *The Boletaceae of North Carolina*).

1974 *The Club and Coral Mushrooms (Clavarias) of the United States and Canada.* New York: Dover Publications (Reprint of the 1923 publication).

1974 *The Gasteromycetes of the Eastern United States and Canada.* New York: Dover Publications (Reprint of the 1928 publication).

# Plants Named for William Chambers Coker

Compiled by William R. Burk, Biology Librarian,
University of North Carolina at Chapel Hill

## Names of Fungi Commemorating William Chambers Coker

(Names in **bold** type are those that have been accepted by one or more botanical authors; synonyms are in *italics*.)

### BASIDIOMYCETES

**Amanita cokeri** (E. J. Gilbert & Kühner) E. J. Gilbert, Iconographia Mycologica 27, Supplement 1 (pt. 2): 372. 1941. *Lepidella cokeri* E. J. Gilbert in E. J. Gilbert & Kühner, Bulletin Trimestriel Société de Mycologie de France 44: 151. 1928; *Aspidella cokeri* (E. J. Gilbert & Kühner) E. J. Gilbert, Iconographia Mycologica 27, Supplement 1 (pt. 1): 79. 1940 & tab. 48. (fig. 6).

**Amanita cokeri** forma **roseotincta** Nagasawa & Hongo, Transactions of the Mycological Society of Japan 25: 373. 1984.

**Amanita cokeriana** Singer, Sydowia 2: 34. 1948.

**Boletus cokeri** H. D. House, *Mycologia* 35: 593. 1943. *Boletus parvulus* Coker & Beers, Boletaceae of North Carolina, University of North Carolina Press, Chapel Hill. 69. 1943, not *B. parvulus* Massee, 1909.

**Clavariadelphus cokeri** Wells & Kempton, Michigan Botanist 7: 46. 1968.

**Clitocybe cokeri** Hesler in A. H. Smith & Hesler, Lloydia 6: 251. 1943.

**Dacrymyces cokeri** McNabb, New Zealand Journal of Botany 11: 475. 1973.

**Entoloma cokeri** Murrill, North American Flora 10: 123. 1917.

**Exidia cokeri** L. S. Olive, Journal of the Elisha Mitchell Scientific Society 60: 18, name only. 1944. Journal of the Elisha Mitchell Scientific Society 74: 41, with Latin diagnosis. 1958.

**Hygrophorus cokeri** A. H. Smith & Hesler, Lloydia 5: 34. 1942. *Hygrophorus gomphidioides* Coker, Journal of the Elisha Mitchell Scientific Society 45: 168. 1929, not *H. gomphidioides* of P. Hennings. 1908.

**Lactarius subvernalis** var. **cokeri** (A. H. Smith and Hesler) Hesler & A. H. Smith, North American Species of *Lactarius*, The University of Michigan Press, Ann Arbor. 144. 1979. *Lactarius cokeri* A. H. Smith and Hesler, Brittonia 14: 425. 1962.

*Lycoperdon cokeri* Demoulin, Lejeunia, n.s. 62: 7. 1972.

**Naematoloma cokerianum** (A. H. Smith and Hesler) G. Guzmán. Mycotaxon 12: 238. 1980. *Psilocybe cokeriana* A. H. Smith and Hesler, Journal of the Elisha Mitchell Scientific Society 62: 193. 1946.

**Phellodon cokeri** Banker, Journal of the Elisha Mitchell Scientific Society 34: 195. 1919.

**Poria cokeri** Murrill, Mycologia 12: 306. 1920.

**Psathyrella cokeri** (Murrill) A. H. Smith, Memoirs of the New York Botanical Garden 24: 405. 1972. *Psilocybe cokeri* Murrill, Mycologia 15: 12. 1923.

**Ramaria cokeri** Petersen. In B. C. Parker, M. K. Roane (eds), The Distributional History of the Southern Appalachians, Part 4. Algae and Fungi, University Press of Virginia, Charlottesville. 291. 1976 [1977].

**Rhizopogon cokeri** A. H. Smith in A. H. Smith and Zeller, Memoirs of the New York Botanical Garden 14 (2): 58. 1966.

**Sebacina** (?) **cokeri** Burt, Annals of the Missouri Botanical Garden 13: 334. 1926.

**Septobasidium cokeri** Couch, Journal of the Elisha Mitchell Scientific Society 51: 40. 1935.

### PLASMODIOPHOROMYCETES

**Sorodiscus cokeri** Goldie-Smith, Journal of the Elisha Mitchell Scientific Society 67: 108. 1951.

### ZYGOMYCETES

**Cokeromyces** Shanor, Mycologia 42: 272. 1950.

**Cokeromyces poitrasii** R. K. Benjamin, Aliso 4: 523. 1960.

**Cokeromyces recurvatus** Poitras, Mycologia 42: 272. 1950.

## *Names of Higher Plants Commemorating William Chambers Coker*

**Ernodea cokeri** Britton in G. B. Shattuck, The Bahama Islands, MacMillan Company, New York. 264. 1905. (Rubiaceae)

**Liatris cokeri** Pyne and Stucky, Sida 14: 205. 1990. (Asteraceae)

**Lycopus cokeri** Ahles ex Sorrie, Castanea 62: 124. 1997. [Published by Ahles by name only in Radford, Ahles, Bell, Guide to the Vascular Flora of the Carolinas, The Book Exchange, University of North Carolina, Chapel Hill. 291. 1964.] (Lamiaceae)

**Malvaviscus cokeri** Britton in G. B. Shattuck, The Bahama Islands, MacMillan Company, New York. 259. 1905. (Malvaceae)

**Torrubia cokeri** Britton. Bulletin of the Torrey Botanical Club 31: 613. 1904. (Nyctaginaceae)

# Notes

LIST OF ABBREVIATIONS

*JEMSS*   *Journal of the Elisha Mitchell Scientific Society*

SCL     South Caroliniana Library, University of South Carolina, Columbia, South Carolina.

SHC    Southern Historical Collection, W. C. Coker papers, #3220, University of North Carolina at Chapel Hill, Chapel Hill, North Carolina. The dates and names of the correspondents mentioned in the text will usually be sufficient guides to the relevant folders in the W. C. Coker papers. If the folders are labeled other than chronologically, this is noted.

PROLOGUE: *The Personality of William Chambers Coker*

1. W. C. Coker, "Science Teaching, Presidential Address before the North Carolina Academy of Science at Wake Forest College, April 28, 1910." Reprint from *North Carolina High School Bulletin* for April, 1910: 20.
2. Coker, like Frederick Law Olmsted, had no formal training in landscape design. Rather, as was also true of Olmsted, his taste for the outdoor life, his keen observation, and his aesthetic sense led to his interest in laying out gardens with trees and shrubs.
3. *News and Observer* [Raleigh, N.C.], August 26, 1944: 6.
4. See chapter 6 for details on the development of this plan.
5. William Chambers Coker Papers #3220, Southern Historical Collection, Wilson Library, University of North Carolina at Chapel Hill. Hereafter, this collection is cited as SHC.
6. *Books from Chapel Hill: A Complete Catalog of Publications from the University of North Carolina Press 1922–1997* (Chapel Hill: U of North Carolina P) xi, 48.
7. Archibald Henderson, *The Campus of the First State University* (Chapel Hill: U of North Carolina P, 1949) 267. William Chambers Coker contributed the greater part of chapter 25 of this book.
8. Family anecdote recounted to me by my mother, May R. Coker.
9. The last two quotations are from audiotaped reminiscences by Dr. Paul Titman of Chicago, December 1998. The audiotape and the transcription, as corrected by Dr. Titman, are now in the author's personal collection.

10. Orange County Register of Deeds, Book of Deeds 59, page 154.

11. Madry, 6.

12. Ibid.

13. Letter of Preston Fox to the author, June 1999. Author's collection.

14. *Recollections of the Major: James Lide Coker, 1837–1918* (Hartsville, South Carolina: Hartsville Museum, 1997) 21.

15. W. C. Coker and Eleanor Hoffman, *Design and Improvement of School Grounds* (Chapel Hill: U of North Carolina P, 1921). See also the story of York, *Recollections of the Major*, 10–11.

16. Letter dated August 23, 1918, from Coker to his brother James. SHC.

17. Dr. Gertrude Burlingham, with whom Coker corresponded about the fungi *Lactariae* from October 1915 through May 1918, was a respected teacher of botany in a Brooklyn high school. See twelve letters of their correspondence, SHC.

18. William Joslin, chairman of the Graham campaign for U.S. Senate for Wake County in 1950, received his contributions.

19. SHC.

20. Coker College archives.

21. SHC. See chapter 2.

22. Proceedings of the Twenty-Third Annual Meeting of the North Carolina Academy of Science Held at North Carolina State College, May 2 and 3, 1924, (*JEMSS* 40:3–4 [December 1924]): 99–100. Other members of the Resolutions Committee at this session of the Academy were Collier Cobb and Lula G. Winston.

23. Proceedings of the Twenty-Fifth Annual Meeting of the North Carolina Academy of Sciences Held at Wake Forest College, April 30 and May 1, 1926 (*JEMSS* 42:1–2): 7. The resolution read as follows: "The North Carolina Academy of Science desires to reiterate that if the present rate of progress and enlightenment in the State of North Carolina is to be maintained and advanced, it is absolutely and unqualifiedly necessary that all those hypotheses, theories, laws and facts which constitute the legitimate content of any field of study may be dealt with at anytime by any teachers. The Academy goes on record as endorsing most emphatically the stand of Dr. H. W. Chase and Dr. W. L. Poteat on the freedom of thought and teaching." Chase, president of the University of North Carolina, and Poteat, president of Wake Forest College, were the chief spokesmen for academia in the fight against the anti-evolutionists.

24. SHC.

25. See letter from C. B. Griffin, July 7, 1916. SHC. Chapter 7 touches on Coker's concern for the community of Chapel Hill.

26. John Archibald McKay was Coker's student assistant for the academic year 1910–11. See University catalog in the North Carolina Collection. Letter of McKay is dated January 28, 1925. The article in question was by Oscar Riddle, "Complete Sex-Transformation in Adult Animals," *The New Republic* 41.529 (January 21, 1925): 225–228. Coker's reply is dated February 10, 1925. SHC.

27. James R. Troyer, *Nature's Champion: B. W. Wells, Tar Heel Ecologist.* (Chapel Hill: U of North Carolina P, 1993) 30–31.

28. See Orrin H. Pilkey, et al., *The North Carolina Shore and Its Barrier Islands: Restless Ribbons of Sand* (Durham: Duke UP, 1998) 116, fig. 6.4.

29. Menu: Soup–cream of tomato, pig salsify, rice, candied sweets, Irish potatoes, corn bread, biscuits, salad, coffee with hot [crullers] and cheese, cordial; guests: McNiders, Greenlaws, Masons, Wilsons, Klutts, Woolens, [Vernons], Howell and Miss B, Whitfields, Nellie R. SHC.

30. Author's conversation with Albert and Laurie Radford, December 2, 1998. Laurie added that during the summer of 1937, which she spent with the Cokers at their Highlands home collecting and photographing mushrooms, Dr. Coker always saw to it that she had milk when drinks were served. It seems to have been generally observed that he was fond of and perhaps needed stimulation. Another of his students, a member of his Introductory Botany class of 1938, said that the students would take careful notice and bet on whether Dr. Coker seemed to have had a nip before his lecture.

31. Coker's correspondents included Donald Culross Peattie, who frequently asked for his advice, as well as occasional correspondence with David Fairchild and Henry Fairfield Osborn, well-known naturalists of his day. As examples of correspondence with Fairchild and Osborn respectively, see letters dated November 24, 1916, and April 18, 1916. On May 27, 1943, Joseph Albers, a well-known artist then teaching at Black Mountain College in Black Mountain, North Carolina, wrote Coker requesting information on medicinal herbs and seeds. SHC.

32. For examples of Coker's correspondence with Dr. Charles S. Sargent, see letters dated May 6, 1918, and January 31, 1919. SHC.

33. Letter to author from Mrs. Preston Fox, June 1999, author's personal collection.

34. See comments on this second-draft letter in chapter 2.

35. "The Distribution of Venus's Fly Trap (*Dionaea muscipula*)," *JEMSS* 43 (1928): 226. Coker's former student C. Ritchie Bell later found a station of this plant near Charleston, S. C., farther south than formerly recorded. See drawing, Coker College archives.

36. Letter written October 28, 1943, and addressed to the editor of the *Chapel Hill Weekly*, SHC.

37. *Recollections of the Major*, 82. See also chapter 1.

38. Preston Fox letter, June 1999, author's personal collection.

39. Laurie Stewart Radford, "The History of the Herbarium of the University of North Carolina at Chapel Hill, NC, 1908–1998." Unpublished essay (copyright applied for 1998): 14.

40. Rhodes Markham was Coker's helper and gardener at "The Rocks" for many years. Letter to Elizabeth Boatwright Coker, September 5, 1934. SHC.

41. See letters to Ralph Sargent, June 25, 1943, and to H. R. Totten, June 16, 1943. SHC.

42. Audiotaped reminiscences of Paul Titman, December 1998, author's collection.

43. Letter from Preston Fox to author, June 1999, author's collection.

44. Author's conversation with Willam Burk, librarian at Couch Library, UNC–Chapel Hill, 1999.

45. Book of Wills, 14, p. 128, Office of the Clerk of Superior Court, Orange County, N.C.

46. *Recollections of the Major*, 96.

47. See reverse of letter dated March 8, 1919, to W. C. Coker from J. L. Coker Jr. SHC.


CHAPTER ONE: *The Student, Early Life*

1. George Lee Simpson Jr., *The Cokers of Carolina: A Social Biography of a Family* (Chapel Hill: U of North Carolina P, 1956) 52–53.

2. Simpson, 86.

3. Hannah Lide Coker, *A Story of the Confederate War: Written at the Request of Her Children, Grandchildren and Many Friends* (Charleston: privately printed 1887; reprint Charlotte, N.C.: The Observer Printing House, 1945) 40.

4. H. L. Coker, 5–7. See also Simpson, 69–78.

5. Simpson, 107–108.

6. Simpson, 49. Simpson is quoting an unpublished manuscript of James Lide Coker.

7. John N. Couch and Velma D. Matthews, "William Chambers Coker," *Mycologia*. 46.3 (May–June 1954): 372.

8. *Recollections of the Major*, 82.

9. Letter dated January 27, 1943, to Miss Mabel L. Pollitzer of the Charleston Museum.

10. Author's conversation with W. C. Coker.

11. Couch and Matthews, 372–73.

12. SHC.

13. Letter dated January 19, 1928, to Dr. C. P. Korstian, Appalachian Forest Experiment Station, Asheville, N. C.

14. Couch and Matthews, 374.

15. To get to his office at the bank from his lodging at 219 Walnut Street he could have walked two blocks toward the river to Front Street to catch the trolley. More likely, the young man walked the three blocks south to his office in the elegant, four-storied bank building on the corner of Front and Princess Streets. See *Wilmington, N.C. City Directory, 1897* (Wilmington: J. H. Hill Printing Co., 1897) 134, for Coker's address in Wilmington. For Wilmington of the 1890s, see Robert Martin Fales, *Wilmington Yesteryear*, ed. Diane Cashman (Wilmington, N.C.: Lower Cape Fear Historical Society, 1984) 34.

16. Couch and Matthews, 374.

17. These letters of recommendation are dated September 27, 1897. I thank Ms. Jennifer Allain Rallo, assistant archivist of the Milton S. Eisenhower Library at the Johns Hopkins University, for sending me copies of these letters.

18. Reminiscences of a niece, Preston Fox, who lived with the Cokers for a considerable time, letter dated June 1999, author's personal collection.

19. Letter from W. C. Coker to Dr. R. P. Cowles, May 21, 1943. SHC. Dr. William Keith Brooks (1848–1908) is known for his morphological studies of floating marine creatures, particularly tunicates (especially *Salpa*) and coelenterates. An artist as well as a scientist, Brooks made fine drawings that are preserved in

the Brooks papers, ms. 66 in Special Collections at the Milton S. Eisenhower Library at the Johns Hopkins University in Baltimore.

20. W. C. Coker, *On The Gametophytes and Embryo of Taxodium*, diss. Johns Hopkins U, 1901. Contributions from the Botanical Laboratory of the Johns Hopkins University, no. 1. Reprint in *Botanical Gazette*, XXXVI (1903) 1–140.

21. Couch and Matthews, 374.

22. W. C. Coker spoke at the biological conference on Friday, October 22, 1926. See the program for the fiftieth anniversary celebration and also the list of donors to the half century fund in the "Annual Report of the President of the Johns Hopkins University" for 1926, Ferdinand Hamburger Jr. Archives, the Milton S. Eisenhower Library of the Johns Hopkins University. See also letter to Coker from the president of the university, dated June 8, 1926, thanking him for his "generous contribution to the Half-Century Fund." SHC.

23. Oral communication to the author from an older member of the Coker family.

24. See Wilhelm Barthlott, "Geschichte des Botanischen Gartens der Universität Bonn" (Bonn: Veroffentlichungen des Stadtarchivs Bonn; Band 48) 1990. See also Wolgram Lobin, Stefan Giefer, Robert W. Krapp, Thomas Pauls, und Wilhelm Bartlott, *Bäume und Sträucher im Botanischen Garten der Universität* (Bonn: Botanischer Garten Universiät Bonn) 1998. I thank Prof. Dr. Wilhelm Barthlott, director of the botanical garden at Bonn, who sent me these publications.

25. Jennie, herself a poet, records in her diary being shown through Goethe's house and sitting at the desk at which he wrote *Faust* and *Wilhelm Meister*. Susanne Gay Linville, *Jennie: Jennie Coker Gay* (Privately printed, 1983) 12.

26. "On the spores of Certain Coniferae," *Botanical Gazette* 38 (September, 1904): 206–13.

27. Other young professors who arrived at UNC–Chapel Hill about this time were Louis Round Wilson, Charles Raper, William Bernard, and C. Alphonso Smith. See William D. Snider, *Light on the Hill: A History of the University of North Carolina at Chapel Hill* (Chapel Hill: U of North Carolina P, 1992) 142–43.

28. After he had become head of the department and a distinguished zoologist, Dr. Wilson was referred to as "Froggy" by students at Chapel Hill.

29. Dr. Wilson was instrumental in getting federal support for the Beaufort laboratory in 1899. The facility became permanent the following year. I am grateful to Douglas A. Wolfe, historian of the Beaufort laboratory, for his letter dated December 12, 2000, providing me with this information on Wilson and Coker.

30. This letter, dated January 28, 1902, was addressed to "Dr. W. C. Coker, Universität, Bonn, Germany." Folder 0, Box 1, W. C. Coker papers, SHC.

CHAPTER TWO: *The Mycologist*

1. Donald P. Rogers, *A Brief History of Mycology in North America*, reprinted in augmented form by the Mycological Society of America (Amherst, Mass.: Newell; Cambridge, Mass.: Harvard U. P. 1981) 33. See list of Coker's publications here appended.

2. Published as an abstract in *JEMSS* 23.2 (June 1907): 48 and in the *Botanical Gazette* 45 (1908): 194–96.

3. *JEMSS* 33: 1–2 (June 1917): 1–88, pls 1–69.

4. *JEMSS* 35: 3–4 (October 1920): 113–82, pls 23, 30–67.

5. *JEMSS* 36: 3–4 (February 1921): 146–96, pls 14–35.

6. Nine reviews and abstracts of this book were published in 1928 and 1929. The *Gasteromycetes* was reprinted by two different publishers, in 1969 and 1974. See William R. Burk, "John Nathaniel Couch (1896–1986), His Contributions to the *Journal of the Elisha Mitchell Scientific Society* and His Scientific Publications," *JEMSS* 116.4 (2000): 302–3.

7. Rogers, *Brief History*, 33.

8. *North American Flora* 2, pt. 1, (1937): 1–76. The article on *Saprolegniales* is with Velma Matthews; the bibliography is with J. H. Barnhart, 69–76.

9. W. C. Coker, "The Gasteromycetes of Venezuela," in Carlos E. Chardon, *Mycological Explorations of Venezuela*, 1934, ch. 20.

10. See Box 42 of 1978 additions to Collection #3220, SHC. Also see a small notebook labeled "Exchange List," which is mostly alphabetized and which contains names and addresses of those with whom he exchanged specimens. Additions of June 1979, Box 2 of 4.

11. Rogers, *Brief History*, 33.

12. *Harvard University Herbaria. The Farlow Herbarium.* 3 August 2002 <http://blodeuwedd.huh.harvard.edu/collections/farlow.htm>.

13. Sixteen of these letters are in the Southern Historical Collection at Chapel Hill; four are in the South Caroliniana Library in Columbia. Farlow wrote all of his letters in an elegant hand.

14. SHC.

15. "The Collybias of North Carolina," *JEMSS* 37 (December 1921): 83–107, pls 1, 4–23; "The Lactarias and Clitocybes of North Carolina," *JEMSS* 38 (September 1922): 98–126, pls 1 (frontispiece) and 7–33; "The *Mycenas* of North Carolina," *JEMSS* 40: 1–2 (August 1924): 49–91, pls 6–30.

16. This species was published as *Exidia cokeri* Olive in *JEMSS* 60 (August 1944): 18–20.

17. For examples of correspondence with Kauffman, see letters of May 30, 1915; October 9, 1915; February 29, 1916; August 6, 1918; and April 3, 1919. SHC.

18. These include *Aleurodiscus, Amanita, Cyphella, Eicheriella, Exidia, Hymenochaete, Peniophora, Polypharaceae, Sebacina, Solenia, Stereum,* and *Thelophora.*

19. SHC.

20. For a photograph of Atkinson with the faculty at Chapel Hill in 1887, see William S. Powell, *The First State University: a Pictorial History of the University of North Carolina*, 3rd ed. (Chapel Hill: U of North Carolina P) 109.

21. George F. Atkinson, *Studies of American Fungi, Mushrooms, Edible, Poisonous, etc.* (Ithaca: Andrews & Church, 1900; 2nd ed., 1901). See Couch and Matthews, 375.

22. William Chambers Coker collection, SCL.

23. SHC.

24. April 18, 1918. SHC.

25. August 25, 1918. SHC.

26. For an account of Atkinson's visit to Chapel Hill, see his letter to his colleague

Professor Whetzel dated September 7, 1918, and preserved in the W. C. Coker papers, SHC.

27. SHC.
28. See letter of November 21, 1918. SHC.
29. All in SHC except for a letter dated February 13, 1919, in the SCL, Columbia.
30. L. R. Hesler, "Biographical Sketches of Deceased North American Mycologists Including a Few European Mycologists," Typescript, January 1975. Couch Library, Coker Hall, UNC–Chapel Hill, N.C.
31. SHC. Coker seemingly noticed in his friends traits characteristic of himself. See chapter 9 for Coker's obituaries of fellow botanists.
32. SHC.
33. SHC.
34. SHC.
35. Letters between February 11 and April 30, 1920. SHC.
36. *JEMSS* 34 (June 1918): 1–62, pls 1–40.
37. Letter dated March 10, 1917, SHC.
38. SHC.
39. SHC.
40. This letter in French is dated February 24, 1920. In May of 1920, Coker sent Bonnier some of his publications, together with a list of all of his publications to date. He mentions Totten's appreciation of Bonnier's friendship while in France after the war. These letters are in the Coker papers, SHC.
41. See letter of March 23, 1932. SHC.
42. SHC.
43. SHC.
44. *JEMSS* 43 (December [?] 1927): 146–51.
45. October 20, 1927. SHC.
46. See letter to Bresadola, March 11, 1922.

CHAPTER THREE: *The Field Botanist*

1. Ralph M. Sargent, *Biology in the Blue Ridge, Fifty Years of the Highlands Biological Station, 1927–1977* (Highlands, North Carolina: The Highlands Biological Foundation, Inc., 1977) 23.
2. See chapter 5 for more about this expedition.
3. Transcribed audiotaped reminiscences of Paul Titman, December 1998, author's personal collection.
4. Audiotaped reminiscences of Paul Titman, December 1998.
5. Author's conversation with Dr. Radford, December 2, 1998.
6. *The Plant Life of Hartsville, S.C.,* (Columbia, S.C.: The State Co., Printers, 1912).
7. *The Plant Life of Hartsville,* 4–11.
8. "The Seedlings of the Live Oak and White Oak," *JEMSS* 28.1 (May 1912): 34–41.
9. See correspondence between Coker and George N. Pindar of the New York Natural History Museum, September 16 and September 20, 1917. SHC.
10. For Coker's own report on such experiments, see Archibald Henderson, *The*

*Campus of the First State University* (Chapel Hill: U of North Carolina P, 1949) 263, 264.

11. Author of *Natural History and Antiquities of Selborne* (1789).

12. "The Distribution of *Rhododendron catawbiense*, with Remarks on a New Form," *JEMSS* 35.1–2 (October 1919): 78.

13. The spelling of the common name for *Dionaea* is not consistent. One sees Venus' Fly trap, Venus Flytrap, venus fly trap, Venus flytrap, and venus flytrap. I choose to use the last form when not quoting another writer.

14. This text is preserved in the SHC.

15. "The Distribution of Venus's Fly Trap (*Dionaea muscipula*)," *JEMSS* 43.3–4 (July 1928): 226.

16. "A Remarkable New Rhododendron," *JEMSS* 51 (1935): 190.

17. *Flora Boreali–Americana*, 1803.

18. Professor of mathematics and astronomy at the University of South Carolina, cousin of W. C. Coker.

19. "*Magnolia cordata* Michaux," *JEMSS* 59.1 (July 1943): 81–88, pls 17–20 and six text figures.

20. "The Distribution of Venus's Fly Trap."

21. Last rhododendron letter of Coker to Holland, July 30, 1919. SHC.

22. SHC.

23. The fruit of the Ogeechee lime is red and quite visible.

24. SHC.

25. "The Woody Smilaxes of the United States." *JEMSS* 60.1 (August 1944): 27–69, pls 9–39.

26. As examples of the Coker-Lunz correspondence on *Smilax*, see letters of February 10, March 8, and March 11 of 1943. SHC.

27. See Coker-Allard letters dated May 14, 1943; May 18, 1943; June 2, 1943; and June 4, 1943. SHC. Currently, one may check the first edition of Catesby in Raleigh, Durham, and Chapel Hill, N.C. Both UNC–Chapel Hill and Duke University possess copies. North Carolina State University possesses this edition on microform.

28. See Coker's letter of thanks to Miss Mabel L. Pollitzer, dated January 27, 1943, for informing him of his appointment as honorary curator of the Charleston Museum. SHC.

29. *The Boletaceae of North Carolina*, 1943, and *The Stipitate Hydnums of the Eastern United States*, 1951.

30. "A Visit to the Grave of Thomas Walter," *JEMSS* 26.1 (April 1910): 31–42, pls 13–14. Coker notes that Walter's book, *Flora Caroliniana*, was published in London in 1788 at the expense of the botanist John Fraser.

31. Coker, "A Visit to the Grave of Thomas Walter," 38.

32. Letter of Judge Henry A. M. Smith in the *Charleston News and Courier* of August 23, 1905.

33. "The Garden of André Michaux," *JEMSS* 27.2 (July 1911) 65–72, pls 1, 2 (p. 80).

34. See his letter to Mrs. H. W. Ravenel dated March 20, 1915, and to Ravenel's daughter dated May 18, 1915, and correspondence dated January 30, 1919, and March

24, 1919, with C. L. Shear of the Bureau of Plant Industry in Washington about Moses Curtis. See folder O of SHC for Coker-Porcher letters about Walter.

35. Author's conversation with Coker's student, Dr. Albert Radford. See catalogs of the University of North Carolina for offerings in botany, North Carolina Collection, Wilson Library. In designing this course, Coker was probably influenced by his course in the history of botany with Professor Noll at Bonn University (November 1901 to April 1902). See his registration book (*Anmeldebuch*) for his courses at Bonn, Coker College archives.

36. See UNC catalogs for 1928–29 and 1929–30, North Carolina Collection.

37. Coker wrote several letters during September and October of 1919 in an effort to learn more about Dr. Gerald McCarthy. He finally received a copy of a letter from Mrs. McCarthy to Dr. John Barnhart of Bronx Park, which gave some information about his later life. As seen in a letter dated November 2, 1920, to Mrs. Jacques Busby in New York, Coker was still trying to learn more about McCarthy. SHC. There is now a book on Gerald McCarthy: Elizabeth M. Ehrenfeld, *Gerald McCarthy, Botanist* (Round Pond, Maine: Road House, 1998). I thank Mr. William Burk for informing me of this publication.

CHAPTER FOUR: *Founder of the UNC Herbarium*

1. Laurie Stewart Radford, "The History of the Herbarium," 1.

2. See the introduction to *The Clavarias of the United States and Canada* (Chapel Hill: U of North Carolina P, 1923) and the Preface of his book, with Alma Holland Beers, *The Stipitate Hydnums of the Eastern United States* (Chapel Hill: U of North Carolina P, 1951).

3. Lewis David de Schweinitz (1780–1835), Moravian theologian who lived in Salem, North Carolina, from 1812 to 1821, was an early collector of fungi in North Carolina. See McVaugh, McVaugh, and Ayers, *Chapel Hill and Elisha Mitchell the Botanist*, 6.

4. See Coker's letter to Norman R. Foerster, Oxford, England, dated June 8, 1921.

5. See letter to Coker from Sigtuna of the botanical institute at Upsala, July 29, 1921. SHC.

6. See chapter 2.

7. See W. C. Coker's letters to James Coker, dated February 2, 1916, and May 19, 1917; letter to W. C. Coker from D. R. Coker, dated October 24, 1927; letter to W. C. Coker from C. W. Coker, dated October 25, 1927. SHC.

8. Coker and Totten, *Trees of the Southeastern States*, 3rd ed. (Chapel Hill: U of North Carolina P, 1945) 108–9.

9. Coker reports finding "on a high dune 8 miles north of Myrtle Beach, S. C. . . . a fine clump of this odd tree, the largest about 11 ½ inches in diameter two feet from the ground." Coker and Totten, *Trees of the Southeastern States*, 3rd ed., 270.

10. Remembrance of author as a passenger in the car.

11. The information in this section comes from letters preserved in the Coker collection, #3220 in the Southern Historical Collection, Wilson Library, UNC–Chapel Hill.

12. See Coker's tribute to Ashe in chapter 8.

13. On March 25, 1932, Dr. Harbison told Coker that he had so little anticipated the loss of his close friend Ashe that he had mailed him a collection of plants on the very day of his funeral. SHC.

14. The little structure that contained Ashe's Raleigh collection still exists. Once a part of the property of Elmwood in Raleigh, the historic home, built in 1813, where Ashe grew up, its present address is 606 Willard Place. I am grateful to Judge George Bason, Ashe's nephew, for giving me this information in a telephone conversation on November 29, 2000, and to Mrs. Doris Proctor Bason for personally showing me this house, Ashe's erstwhile herbarium storehouse in Raleigh.

15. See below the letter of W. C. Coker to President Graham, dated September 24, 1932. SHC.

16. Letter to Coker from Harbison, dated March 25, 1932. This letter and those that follow about the Ashe herbarium are in the Coker papers of the Southern Historical Collection in Wilson Library at the University of North Carolina.

17. Harbison did not want to separate the "types" from the herbarium. See letter to Coker from Harbison in Highlands of April 4, 1932. SHC. The word "cotype" (see below) probably meant for Coker and Totten specimens gathered at the same time and place as Ashe's "type." These terms are no longer used in the International Code of Botanical Nomenclature (St. Louis, 2000). Dr. Rogers Mc-Vaugh, of the Department of Botany, UNC at Chapel Hill, kindly provided me with this information at the request of Mr. William Burk, librarian at the Couch Library at Coker Hall, UNC, in a memo dated July 27, 2001, and enclosed in Burk's letter to me dated July 30, 2001. Related contemporary terms are explained in Dr. McVaugh's memo in too great length for inclusion here.

18. As examples, see letters between Coker and Harbison of March and April of 1932. SHC.

19. Letter from Harbison to Coker dated December 5, 1932. Apparently Harbison had a role in the preservation of some neglected specimens of Elliott's herbarium. His exact role is not certain. A decade later, Coker also played a very small part in the preservation of the remnant of Elliott's collection in the Charleston Museum. G. Robert Lunz, a member of the three-person directing committee of the Charleston Museum in the absence of E. Milby Burton during World War II, wrote to Coker on April 22, 1943, that they had decided to use his gift to the museum for a case to house the Elliott herbarium. He asked for suggestions as to how the case should be constructed. Coker replied on April 27 with precise instructions for the design of the cabinet. On May 11, 1943, Lunz wrote Coker that materials had been ordered and they hoped to start the construction the following week. SHC. For the role of Lunz at the museum during the war years, see Sanders and Anderson, 181. I am indebted to Ms. Sharon Bennett, archivist at the Charleston Museum, for her help with information on the history of the Elliott herbarium and Coker's interest in the museum. See Sanders and Anderson, 39 and 123, for information on the Elliott herbarium.

20. See n. 17 above for definition of "type" and "cotype" as used in this paragraph.

21. B. W. Wells, *The Natural Gardens of North Carolina: with Keys and Descriptions of the Herbaceous Wild Flowers Found Therein* (Chapel Hill: U of North Carolina P, 1932). In this letter, Coker wanted Harbison's suggestions and criticism about Wells's book. He says, "I think some of his writings about the mountains are considerably too bookish. He talks about several of the various scarcest things in the state as if one would be apt to run across them in walking about in the mountains, for instance, *Celastras*." SHC.

22. Plants were dipped in insecticide for protection before mounting.

23. Letter from Harbison to W. C. Coker dated April 20, 1934. SHC.

24. See David R. Coker papers, South Caroliniana Library, University of South Carolina, for January 16, January 19, and February 20 of 1935.

25. L. S. Radford, "History of the Herbarium," 8. (Language slightly changed for clarity.)

26. Author's conversation with Laurie Stewart Radford, December 2, 1998.

27. L. S. Radford, "History," 16.

28. L. S. Radford, 19.

29. L. S. Radford, 15.

30. L. S. Radford, 16 and 20.

31. Coker had proposed a "collection of trees and shrubs native to North Carolina" in a letter to President Chase, 1927. See *Master Plan for the North Carolina Botanical Garden, A Guide for Development*, 5. Again he stated this proposal in an unpublished paper of 1944. L. S. Radford, 29. See final chapter, on Coker's legacy.

32. There are two notebooks in the unprocessed addition of 1979 to the Coker papers that contain careful notes on shrub garden plots. The first describes plots 1–177 and the second, plots beginning with 178. SHC.

33. L. S. Radford, 28–29.

34. Radford, Ahles, and Bell, v.

35. Mrs. Mary Felton was his able assistant in the poisoning and mounting of specimens.

36. Correspondence of January 2001 between Dr. Massey and the author.

37. See Carol Ann McCormick, "Business Usual and Unusual at the UNC Herbarium," *North Carolina Botanical Garden Newsletter* 29.4 (July–August 2001) 1.

38. *North Carolina Botanical Garden Newsletter* 28.4, (July–August, 2000).

39. See note 31, above. For some of the information in these concluding paragraphs, I am indebted to Dr. Peter White and to a news sheet of the Friends of the UNC Herbarium.

CHAPTER FIVE: *Champion of the Highlands Biological Station*

1. See Ralph M. Sargent, *Biology in the Blue Ridge: Fifty Years of the Highlands Biological Station, 1927–1977* (Highlands, North Carolina: The Highlands Biological Foundation, Inc., 1977) 2–3. Dr. Robert Wyatt, director of the Station, offered me access to information on Highlands and the work there. He kindly sent me information from a proposal, which has resulted in a grant from the National

Science Foundation. This information can also be read in the "2002 Annual Announcement" of the Station. Mr. Randolph P. Shaffner sent me a copy of his most helpful book, *Heart of the Blue Ridge: Highlands, North Carolina,* as soon as it was published.

2. April 23, 1932, letter of Coker to Warren Weaver, director general of education for the Rockefeller Foundation. Collection 3220, Series E, Folder 847, SHC.

3. Randolph P. Shaffner, *Heart of the Blue Ridge: Highlands, North Carolina* (Highlands, North Carolina: Faraway, 2001) 343.

4. Sargent, 7, 8.

5. Sargent, 9.

6. Sargent, 12, 13.

7. In 1931, Dr. Coker had purchased Lots 47 and 48 on Linwood Lake, now Lake Ravenel, near which are located both the Station and Coker's former cottage. His home is on what was Lot 47. The Coker Rhododendron Trail, deeded to the Highlands Biological Station by Mrs. Coker in 1954, is part of this property, crossing what was Lot 48. In 1937, Dr. Coker purchased Lot 22, on which the Coker Laboratory and the Thelma Howell Administration Building for the Station were later constructed. These land transfers are recorded in Book J-4, Office of the Macon County Register of Deeds, Franklin, North Carolina.

8. Shaffner, 396, 454.

9. David R. Coker Papers, South Caroliniana Library, University of South Carolina, Columbia, S.C.

10. "Mycological Society of America: Summer Foray," *Mycologia* 25 (1933): 550–52.

11. He included in his petition the preservation of the southern third of Smith Island on the coast of North Carolina. See letters to Senators Bailey and Reynolds dated February 10, 1936, and letter to Dr. C. L. Shear, Bureau of Plant Industry, April 9, 1936. W. C. Coker papers, Series E, Highlands, folders 848 and 849. SHC.

12. Letter dated April 26, 1939 to Daniel C. Roper, U.S. Secretary of Commerce, who was the father of his sister-in-law, Mrs. David Coker.

13. Letter from Coker to Whitney, February 5, 1942, in reply to Whitney's letter of the previous December. SHC Series E, Highlands papers, folder 854.

14. Sargent, 15. Coker had been vice president of the Station in 1931 and 1932. Sargent, 142.

15. Sargent, 24.

16. Sargent, 142–43.

17. SHC.

18. See chapter 4.

19. Dr. Olive and Dr. John N. Couch were the only two botanists of the Coker years at Chapel Hill who were elected to membership in the National Academy of Sciences. Dr. Coker himself was not a member. Dr. Eugene Odum spent a summer studying birds of the area. Later a professor at the University of Georgia, Odum had grown up in Chapel Hill, where his father was professor of sociology.

20. See "2002 Annual Announcement of the Highlands Biological Station" and note 1 above.

21. Sargent, 56.

22. Sargent, 59.

23. Coker reported during the next two years failure of a committee of biologists from UNC, Vanderbilt, and Duke Universities to achieve affiliation of the laboratory with these institutions. Sargent, 21.

24. Letter to W. C. Coker from E. E. Reinke, March 12, 1941. Coker replied on March 20, 1941, that "it begins to appear that it would be better for the laboratory to be owned by one institution rather than three."

25. Sargent, 133.

26. "The Highlands Biological Station, 2001 Annual Announcement," Highlands, North Carolina: Office of Public Information/Publications, November 2000: 1, 16.

27. See C-5, Book J-4, 289 of the Macon County Office of the Register of Deeds, Franklin, North Carolina.

28. Sargent, 20.

29. W. C. Coker Collection, #3220, SHC, Series E, Highlands Papers, folder 843. September 26, 1941.

30. Treasurer's Report to the Annual Meeting, August 26, 1944. W. C. Coker collection, Series E, Highlands papers, folder 856.

31. Highlands Series E, folder 856. SHC.

32. Highlands Series E, folder 852. SHC.

33. For the papers based wholly or in part on research done at the Highlands Biological Station during the first fifty years of its existence, see the list in Sargent, 144–56.

34. See proposal #DBI-0121490 submitted to the National Science Foundation by the Highlands Biological Station entitled "New Housing for Researchers, Faculty, and Students at Highlands Biological Station," section 4.8.

CHAPTER SIX: *Founder of the Arboretum, Planner for the UNC Campus*

1. Archibald Henderson, *The Campus of the First State University* (Chapel Hill: U of North Carolina P, 1949) 269. See also the Harris map of 1795 between pages 8 and 9 of Henderson for the original plan for "ornamental grounds." The walk across the wet meadow is now the central straight path running the length of the Arboretum, long called "President's Walk." See also *Carolina Alumni Review*, November 1943, 68.

2. Curtis Brooks, "A Brief History of the Coker Arboretum: 'The Coker Years,'" *North Carolina Botanical Garden Newsletter* 13.5 (September–October, 1985) 1–2. For the sequel, see 13.6 (November–December, 1985) 2–3. For the Arboretum's history until the mid-1980s, I have depended heavily on these two articles by Curtis Brooks.

3. See letter from Caudell to Coker dated October 30, 1909. SHC.

4. Letter dated October 20, 1927, to Professor Preston Edwards, a cousin of Coker's. SHC.

5. Henderson, 258–68.

6. Henderson, 258.

7. Letter to the president and trustees, November 16, 1918. SHC.

8. Henderson, 261.

9. Presumably of the Rockefeller Foundation.

10. Henderson, 259.

11. Coker arrived "at the disappointing conclusion that *cordata* is only a rather vaguely defined, marginal extension of the southern yellow-flowered variety of *M. acuminata.*" *JEMSS* 59.1 (July, 1943) 81–88, pls 17–20, especially pp. 86–87. He had suspected at one point that the two were separate species. See chapter 3.

12. See chapter 8 about opportunities for American military personnel after the armistice of November 1918. See p. 133 for Totten's drug garden.

13. North Carolina pharmacists active today remember Dr. Totten as their teacher. For example, Herman Hallet Daniels, UNC graduate of the School of Pharmacy in 1952, student of Totten in 1948, and currently pharmacist at Capps Drug Co. in Ahoskie, North Carolina, said of his teacher, "Dr. Totten was the most knowledgeable man in plants and botanicals that I have ever known. He was very down to earth, an interesting teacher who caught and held your attention." Conversation between Mr. Daniels and the author in the post office of Winton, North Carolina, on October 27, 2001, at the celebration of Winton's Heritage Day. Robert Cotton, among Totten's last students, presently chief pharmacist at Eckerd's, a drugstore at Five Points in Raleigh, commented on how much Dr. Totten knew and how interesting he made what could have been a dry subject. Conversation with the author at Eckerd's on November 7, 2001.

14. There are two notebooks in the W. C. Coker papers with careful records of garden plots planted with shrubs in the Mason Farm area. The first contains records of plots 1 through 177; the other records plots numbered 178 onward. These notebooks (1916?) contain an alphabetized index by botanical names. Apparently, Coker had intended to write a book on the native shrubs of North Carolina, a project never brought to completion. See unprocessed addition of 1979 to Collection 3220, Box 42, 2 of 4. SHC. Incidentally, Coker donated the land on which the original golf course was constructed, along Laurel Hill Road.

15. This information on the drug garden of the Arboretum is derived from Anders S. Lunde, "The First Drug Garden at the University of North Carolina," *North Carolina Botanical Garden Newsletter* 12. 4 (July–August, 1984) 1–2.

16. Henderson, 259–64.

17. Henderson, 259.

18. A metal plaque on the tree in the Coker Arboretum identifies it. An explanation of its origin is sketched beside it.

19. See International Checklist of Cultivated Ilex, Part 1. *Ilex opaca*, National Arboretum Contribution no. 3, Agricultural Research Service, U.S. Department of Agriculture, March 1973.

20. I am indebted to Mr. William Burk of the Couch Library in Coker Hall, who tracked down the holly named Pearle LeClair and explained the role of Mr. LeClair in the planting of hollies on campus. See William R. Burk, "Of Hollies

and Horticulture: A Brief Look at the Life of Francis Joseph LeClair." *North Carolina Botanical Garden Newsletter* 10.1 (January–February 2002): 10.

21. Robert B. House, "In Warm Memory of Francis J. LeClair," *The Chapel Hill Newspaper*, January 13, 1974, North Carolina Collection Clipping File through 1975, UNC Library, Chapel Hill, 1974, #607.

22. The fact that the committee was referred to as "Buildings and Grounds" after Coker's tenure, a term used in most universities, may reflect a different priority of responsibility or simply a traditional title. Under Coker, the term was "Grounds and Buildings."

23. Elisha Mitchell (1783–1857) arrived at the University in 1818. In 1829, Dr. Mitchell, who had many other duties at the University, was appointed "Superintendent of the property and financial concerns of the University," a position which gave him responsibility for maintenance and improvement of University grounds. In 1838 Mitchell, who came from Connecticut, where he was familiar with the building of stone fences, "entered upon the arduous task of superintending the building of a rock wall entirely around the Campus." See Henderson 60–62, 124 and McVaugh, McVaugh, and Ayers, 3.

24. Mr. Jones later became curator of the Asian Garden at Sarah P. Duke Gardens at Duke University.

25. *North Carolina Botanical Garden Newsletter*, 13.6 (November–December 1985): 3.

26. Mr. Brooks is now town arborist and supervisor of horticultural work for Chapel Hill.

27. Charlotte Jones-Roe of the North Carolina Botanical Garden provided me with information on the curators of the Arboretum in a letter dated October 26, 2001, author's collection.

28. Snider, 153.

29. See Kelly Russell, "In Broad Daylight," *Carolina Alumni Review*, March/April 2001, 56–62. An Arboretum bench in perpetual memory of Suellen Evans is planned.

30. For the significance of this tree for the University of North Carolina at Chapel Hill, see Coker, *Trees of the Southeastern States* (Chapel Hill: U of North Carolina P) 189.

31. Thomas Wolfe, *Passage to England: a Selection*, edited by Suzanne Stutman and John L. Idol Jr. (Rocky Mount, N.C.: Walker–Ross Printing, The Thomas Wolfe Society, 1998) 23, 24. I am indebted to Dr. Idol for this reference.

32. This little building remains today, albeit somewhat enlarged.

33. Richard Walser, *Thomas Wolfe Undergraduate* (Durham, N.C.: Duke University Press, 1977) 28.

34. SHC.

35. Thomas Wolfe, *Look Homeward, Angel* (New York: Scribners, 1929) 330.

36. *The Letters of Thomas Wolfe, Collected and Edited with an Introduction and Explanatory Text by Elizabeth Nowell* (Charles Scribners's Sons, 1956) 8.

37. For this letter and its response, see SHC.

38. This shrub garden was still on the Mason Farm property.

39. Letters mentioned in the preceding five paragraphs are preserved in SHC.

40. These two letters are in the W. C. Coker Collection in the South Caroliniana Library at University of South Carolina. Most of this correspondence between Coker and Nolen is preserved in the Southern Historical Collection in Chapel Hill. Those letters in the South Caroliniana Library and in the archives at Coker College are so indicated.

41. Coker College archives.

42. Coker College archives.

43. SHC.

44. Snider, 169.

45. See letters to Curtis, to Grandgent, and to the architectural firm of McKim, Mead, and White, February 16, 1920. SHC.

46. This letter to J. A. McKay was dated July 30, 1920.

47. See SHC for these letters to Chase of April and June 1920.

48. Letter from McLean dated October 10, 1923. SHC.

CHAPTER SEVEN: *Landscape Designer, Extension Agent*

1. See letter from President Emilie W. McVea of Sweet Briar College, August 3, 1917. On September 14, Coker wrote Miss McVea that he was delayed because of the loss of his assistant to war service. She replied on September 23 that she would have no other person than Coker to do the work and that she would wait for his convenience. SHC.

2. See letter to E. C. Coker, dated January 11, 1925. SHC.

3. See letter to W. C. Coker from John Nolen, June 1, 1915, Coker College archives, and letters between Nolen and Coker dated March 31 and April 16, 1920. SHC.

4. See correspondence with S. W. Garrett, bursar of Coker College, who was responsible for the grounds, January 22 and January 25, 1919. SHC.

5. Among those requesting and receiving his help were David R. Coker and Joseph J. Lawton of Hartsville, his brother and brother-in-law, and his friend Bright Williamson of nearby Darlington. See letters dated January 31 and February 19, 1916; February 17, 1917; and July 5, 1917. SHC.

6. See chapters 4 and 8 for Coker's role in getting the help of T. G. Harbison and C. R. Bell for Kalmia. His former students Velma Matthews and Budd Smith, teachers at Coker College, also helped with the collection of plants.

7. See, for example, the Coker-Tarbox correspondence related to planting at Brookgreen Gardens. Some of these letters are dated October 21, 1943; July 24, 1944; March 28, 1946; and April 29, 1946. SHC. Coker and Tarbox exchanged many letters between September 17, 1937, and May 28, 1946. See chapter 3.

8. On July 24, 1944, Tarbox wrote Coker that he was on the lookout for the ripe fruit of *Smilax pumila*. Coker was then preparing his article "The Woody Smilaxes of the United States" *JEMSS* 60 (1944): 27–69. The Ogeechee lime was another plant that Tarbox sought for Coker. See chapter 3 and letter dated April 22, 1943. SHC.

9. See March 1917 correspondence about landscaping at Carr Mills in Durham; December 5 and 6, 1916, correspondence with Mrs. E. Oscar Randolph, wife of his former student then teaching at Elon College, about improvement of the

local train station; letter of early November 1928 about plans for the Asheville cemetery; and a letter of March 4, 1920, to Mrs. Pruden with a plan for the library grounds at Edenton. SHC.

10. See William D. Snider, *Light on the Hill: A History of the University of North Carolina at Chapel Hill* (Chapel Hill: U of North Carolina P, 1992) 160. In his inaugural address in April of 1915, E. K. Graham said, "The state University is the instrument of democracy for realizing all these high and healthful aspirations of the state. . . . It would conceive of the present state and all of its practical problems as the field of its service. . . . " Louis R. Wilson, *The University of North Carolina, 1900–1930: The Making of a Modern University* (Chapel Hill: U of North Carolina P, 1957) 185.

11. Wilson, 209.

12. Henry Griscom Parsons, "Children's Gardens," *Plant World* 9 (1906): 237–39. Box 21 (unprocessed addition, box 1 of 4). SHC.

13. January 20, 1908. Box 21, (unprocessed addition, box 1 of 4). SHC.

14. See letters dated May 12, 1909 and March 14 and 22, 1910, W. C. Coker collection P/10370, South Caroliniana Library, University of South Carolina in Columbia.

15. On January 12, 1917, he wrote to a Mrs. Bryant that he had had to give up lectures for the extension program, as "work in laying off and planting school grounds is taking too much time." SHC.

16. (With Eleanor Hoffman), *Design and Improvement of School Grounds*, Bureau of Extension Bulletin, Special Series No. 1 (Chapel Hill, N.C., 1921) and (with Enid Matherly) *How to Know and Use the Trees*, Extension Bulletin No. 14 (Chapel Hill, N.C., 1924). SHC.

17. As examples, see the many letters to Coker with requests for the landscaping of North Carolina schools from September 21 through November 14 of 1920. SHC.

18. As examples, see letters to Mrs. Long and Mrs. De Vane, October 6 and November 11, 1920, Coker Collection, SHC.

19. See letter to L. R. Wilson, September 21, 1920, SHC. Miss Hoffman was to be coauthor of his first extension bulletin.

20. SHC.

21. SHC.

22. See letters to the Chapel Hill postmaster, Mr. McRae, of May 15 and June 3, 1920, and a letter dated July 7, 1916, from C. B. Griffin, clerk of the Chapel Hill Board of Aldermen.

23. See letter to Watts Hill dated May 18, 1932, SHC. That Coker's thinly veiled rebuke was directed toward the very person who in 1926 had donated for use on the campus a hothouse and who would in December of 1932 buy the precious Ashe herbarium for the University testifies to the staunch friendship of the two men. See Coker's letters of June 28, 1926, and December 22, 1932. SHC.

24. (With E. O. Randolph), "Observations on the Lawns of Chapel Hill," *JEMSS* 31 (1915): 113–19; "Winter Grasses of Chapel Hill," *JEMSS* 31 (1915): 156–61; "The Lawn Problem in the South," *JEMSS* 31 (1915): 162–65.

25. See *JEMSS* 31, 160–61.

26. *Prunus caroliniana* is the botanical name for Carolina cherry laurel. Mock orange is *Philadelphus*. See Radford, Ahles, and Bell, 509, 520. This 1915 article gives a confusing nomenclature.

27. *JEMSS* 31, 165.

28. See reference to this letter in chapter 6.

29. Coker Papers, SHC. See also Henderson, 277–78.

30. This land was located where the present North and Boundary Streets meet. See Book of Deeds 59, page 154, Orange County Register of Deeds, Hillsborough, N.C.

31. See *North Carolina Botanical Garden Newsletter* 29.1 (January–February 2001): 15.

32. Dr. Coker mentions the row of nineteen hollies that he brought into his garden from the woods in about 1914 (thirty-five years before his comment in 1949). Henderson, 263.

33. Audiotaped reminiscences of Dr. Paul Titman of Chicago, Coker's former student, author's personal collection.

34. Audiotaped reminiscences of Dr. Paul Titman.

CHAPTER EIGHT: *The Teacher and His Students*

1. Letter to Professor R. C. Bently of Urbana, Illinois, dated January 19, 1917. SHC.

2. Letter dated January 27, 1919, to Dr. Albert F. Woods, Washington, D.C. SHC.

3. Letter dated November 12, 1919. SHC.

4. Couch and Matthews, 377.

5. See "Notes on Fungi, with a Description of a New Species of Ditiola," *JEMSS* 46 (1930): 182.

6. Paul Titman (1920–2000), of the class of 1941, earned a Ph.D. from Harvard and became a professor of botany at Illinois State College in Chicago. The text is an excerpt from my transcription of Dr. Titman's taped oral memoirs, recorded at my request. I received his recording by mail on December 24, 1998. Shortly afterward, he read my transcription and returned it to me with minor corrections of my spelling of botanical terms.

7. See Hyman folder at the General Alumni Association records office at Chapel Hill and also a letter dated January 29, 1917, from Coker to the dean of Princeton, recommending Hyman for a fellowship in biology. In another dated April 5, 1917, Hyman thanks Coker for assisting him in obtaining the appointment. SHC. See below in this essay for more on Hyman.

8. Goldston later became an oil producer in Houston, Texas. See Goldston folder, General Alumni Association records office, UNC–Chapel Hill.

9. The three notebooks, those of Hyman, Goldston, and Totten, are to be found in Box 2 of four boxes that constitute the "Unprocessed Addition of 1979" to the W. C. Coker collection, SHC. See below for more about Totten.

10. *Bibliography of North American Flora*, vol. 2, pt. 1, (November 1937) 69–76.

11. These coauthors are H. C. Beardslee, Eleanor Hoffman, Enid Matherly, P. M. Patterson, J. F. Holmes, C. F. Korstian, O. W. Hyman, E. O. Randolph, H. R.

Totten (several), Mary S. Taylor (with credit to Herbert Hechenbleikner), Jane Leitner, Leland Shanor, and Alma Holland Beers. See list of the publications of W. C. Coker below.

12. See first senior photograph of the *Yackety Yack* of 1904 and p. 14 of the 1905 *Yackety Yack*, General Alumni Association library, UNC–Chapel Hill.

13. See "On the Spores of Certain Coniferae," *Botanical Gazette* 38 (September 1904): 206–13, n. 8. Much later, in November of 1920, Coker invited H. A. Allard to address the Elisha Mitchell Society on his work as a plant physiologist at the Bureau of Plant Industry in Washington. Allard declined because of his tendency to experience stage fright while addressing audiences. See letter dated November 15, 1920. SHC.

14. H. R. Totten, A.B. 1913, M.A. 1914, Ph.D. 1923, all in botany under Coker's direction.

15. J. N. Couch, A.B. 1918 (awarded in 1919), M.A. 1922, Ph.D. 1924, all in botany under Coker's direction.

16. See comments on Duncan Starr Johnson in Coker's obituary of his botany professor at Johns Hopkins in chapter 9.

17. See H. R. Totten folder, Class of 1913 twenty-fifth anniversary record questionnaire, General Alumni Association records office, UNC–Chapel Hill.

18. For these letters about Couch's year at Cold Spring Harbor, dated February 24, 1925, and April 18, 1926, see SHC.

19. See Hyman folder, General Alumni Association records office, UNC–Chapel Hill.

20. Patterson was coauthor with Coker of "A New Species of Pythium," *JEMSS* 42: 247–50.

21. Velma D. Matthews is author of "The Ferns and Fern Allies of South Carolina," rpt. from *The American Fern Journal* 30–31.3, 4, and 1, July–September, October–December, 1940, and January–March, 1941, 73–80, 119–28, and 4–12.

22. See Radford, Ahles, and Bell, preface, v.

23. B. E. Smith corresponded with Dr. Coker while teaching in Hartsville and while serving in the navy in World War II. As examples of this correspondence, see letters from Smith to Coker dated December 13, 1936, and February 13, 1943. SHC.

24. These two recollections are from a letter to the author dated January 27, 2001, from Helen Parker Kelman, UNC student in botany during the years 1946–48. Author's collection.

25. Author's conversation with Albert Radford, December 2, 1998.

26. Letter concerning the student loan fund to Coker from Dean of Students Francis F. Bradshaw, March 18, 1932. SHC.

27. Letter from Mrs. Janie Edwards Remsburg of Fayetteville, daughter of his father's sister Anna, dated September 1, 1919. SHC.

28. Letter dated May 23, 1917, to J. I. Somers of Burlington. SHC.

29. Author's conversation with C. Ritchie Bell, December 2, 1998.

30. Perry's letter was dated January 6, 1918. SHC.

31. Letter from E. Oscar Randolph, dated July 18, 1916. SHC.

32. Audiotaped reminiscences of Paul Titman, December 1998. Author's collection.

33. Wiggins insisted that, as Atlantic Coastline Railway chairman, his office be in Hartsville. I am indebted to James C. Fort for this information.

34. See file on Goldston in the General Alumni Association records office at UNC–Chapel Hill.

35. See chapter 4 for more on Harbison's role at Kalmia Gardens.

36. Coker stated on his application for overseas service in the American Red Cross that he wanted to do reconstruction work and cited his experience in horticulture, landscape design, agriculture, plant breeding, and botany. He gave his languages as German and Spanish but stated his preference to work in France. A copy in his own hand of this application form is preserved in the Coker College archives.

37. See, as examples, correspondence with Charles O'Kelley and Budd Smith in October and November of 1943. In a letter dated June 30, 1943, Coker gave Paul Titman news of his fellow students with whom he (Coker) was in contact: Budd Smith, Bill Ziegler, Howard Reid, Roland Fields, Laurie and Al Radford, Lee Adams, and Lindsay Olive. Most of them were in military service.

38. Totten's role in the Arboretum's drug garden is discussed in chapter 6. See also chapter 6 for comments of pharmacists on Totten as their teacher.

39. Subsequent expanded editions came out in 1934 and 1945. See chapters 3 and 6.

40. Coker later studied in Bresadola's herbarium in Stockholm and corresponded with him in Italy. See chapter 2.

41. For copies of these letters of the spring and summer of 1919 between Coker and Totten and Coker and Couch, see SHC.

CHAPTER NINE: *Writer and Editor*

1. *On the Gametophytes and Embryo of Taxodium.* This early botanical dissertation by a student at Johns Hopkins was published by the *Botanical Gazette* at the University of Chicago. See list of Coker's publications here appended.

2. "Vegetation of Bahama Islands," 185–270, pls 1, 33–47 in G. B. Shattuck, ed., *The Bahama Islands* (New York: Macmillan, by permission of the Geographical Society of Baltimore, 1905), xxxii.

3. *JEMSS* 19 (1903): 42–49.

4. *JEMSS* 20 (1904): 35–37.

5. *JEMSS* 23 (1907): 134–36.

6. *JEMSS* 32 (1916): 66–81.

7. Chapel Hill, N.C.: published by W. C. Coker, 1916.

8. "Science Teaching, Presidential Address before the North Carolina Academy of Science at Wake Forest College, April 28, 1910." Reprint from the *North Carolina High School Bulletin* for April 1910, 18.

9. "A Visit to the Yosemite and the Big Trees," *JEMSS* 25.4 (December 1909): 143.

10. *The Plant Life of Hartsville,* 1912, 17 and 24. Much later, in 1944, Coker was to publish an article on smilax, "The Woody Smilaxes of the United States," *JEMSS* 60 (1944): 27–69.

11. Coker, "A Visit to the Grave of Thomas Walter," 33.

12. "William Willard Ashe," Reprint *JEMSS* 48.1 (October, 1932): 40–47. Though this article is signed by two other members of a committee appointed by the North Carolina Academy of Science, the writing has a strong flavor of Coker's style.

13. W. C. Coker, "Lars Romell," *JEMSS* 43 (1927): 146–51.

14. In April of 1928, Dr. Johnson visited W. C. Coker at Chapel Hill. See letter to Coker dated April 1, 1928. SHC.

15. "Professor Duncan Starr Johnson," *Science* 86 (December 3, 1937): 512.

16. Conversation of the author with Albert Radford. Volume 62 (1946) of the *Journal of the Elisha Mitchell Scientific Society* was dedicated to its longtime editor. The inscription on the prefatory page reads as follows: "To William Chambers Coker, Ph.D., LL.D., Kenan Professor Emeritus Of Botany in The University of North Carolina, who for forty-one Years (1904–1945) was Editor-in-Chief of The Journal of the Elisha Mitchell Scientific Society, this Volume is Affectionately Dedicated."

17. See reference to Coker's field trip to duplicate Gray's search in chapter 3.

18. Coker and Totten, *Trees*, 1945, 282.


EPILOGUE: *The Legacy of William Chambers Coker*

1. Coker, *Design and Improvement of School Grounds*, 6.

2. Venus flytrap is protected by a North Carolina law introduced by Representative Mintz of Southport and passed during the 1951 session of the General Assembly. See chapter 367 of session records for that year.

3. *Master Plan for the North Carolina Botanical Garden, A Guide for Development*, (March 1992), 5.

4. "We propose the establishment at Chapel Hill of a really adequate collection of living trees and shrubs of the southeastern states." L. S. Radford, "The History of the Herbarium," 29. For this unpublished manuscript, see the beginning of folder 732 of the W. C. Coker papers, SHC. The document, signed by Coker, is dated only October 1944.

5. Master Plan, i.

6. Velma Matthews, Budd Smith, Arthur William Ziegler, Waring Webb, and Robert McIntyre.

7. See letter of W. deB. McNider, dated June 17, 1920, SHC.

8. See "Vegetation of the Bahama Islands" in *The Bahama Islands* (Baltimore: Geographical Society of Baltimore, 1905) 33–47, 185–270.

9. R. W. Madry, "Beauty of Chapel Hill is Monument to Coker," *The News and Observer*, Raleigh, N. C., Sunday, August 27, 1944, 6.

10. Couch and Matthews, 376.

11. Henderson, 263.

12. Audiotaped reminiscences of Dr. Paul Titman, December 1998. Author's collection.

13. Daniel Chester French to Julia Book, Coker Papers (#3220), Series A, folder 282, SHC.

14. Letter to author dated December 16, 1998, from Claire Freeman of Raleigh. Author's collection.

15. Laurie Stewart Radford, *History of the Herbarium*, 29, 30.

16. Dr. Peter White, director of the North Carolina Botanical Garden, in a letter dated September 28, 2001, directed me to Latham and Ricklefs, who demonstrate that eastern North America has more tree species than northern, central, and eastern Europe. See their article "Continental Comparisons of Temperate-Zone Tree Species Diversity" in E. Ricklefs and Dolph Schluter, *Species Diversity in Ecological Communities: Historical and Geographical Perspectives*, 294-314, especially 311–14.

# Selected Bibliography

*A Backward Glance: Facts of Life in Chapel Hill*. The Chapel Hill Bicentennial Commission, 1994.

Atkinson, George F. *Studies in American Fungi, Mushrooms, Edible, Poisonous, etc.* 2nd ed. Ithaca, N.Y.: Andrews and Church, 1901.

Barthlott, Wilhelm. *Geschichte des Botanischen Gartens der Universität Bonn*. Sonderdruck aus Bonn-Universität in der Stadt hrsg. von Heijo Klein (Veröffentlichungen des Stadtarchivs Bonn; Band 48) 1990.

*Books from Chapel Hill 1922–1997: A Complete Catalog of Publications from the University of North Carolina Press*. Chapel Hill: U of North Carolina P, 1997.

Burk, William R. "John Nathaniel Couch (1896–1986), His Contributions to the *Journal of the Elisha Mitchell Scientific Society* and His Scientific Publications." *Journal of the Elisha Mitchell Scientific Society* 116.4 (2000): 297–306.

Coker, Hannah Lide. *A Story of the Confederate War: Written at the Request of Her Children, Grand-children and Many Friends*. Charleston, S.C., 1887. Charlotte, N.C.: The Observer Printing House, 1945.

Couch, John N., and Velma Matthews, "William Chambers Coker." Reprint from *Mycologia* 46.3 June–July 1954: 372-83.

Ehrenfeld, Elizabeth M. *Gerald McCarthy, Botanist*. Round Pond, Maine: Road House P, 1998.

Elliott, Stephen. *A Sketch of the Botany of South Carolina and Georgia*. Charleston: J. R. Schenck, 1821–24.

Fales, Robert Martin. *Wilmington Yesteryear*. Ed. Diane C. Cushman, sponsored by the Lower Cape Fear Historical Society. Wilmington, N.C.: Robert Martin Fales, 1984.

Henderson, Archibald. *The Campus of the First State University*. Chapel Hill: U of North Carolina P, 1949.

Hesler, L. R., "Bibliographical Sketches of Deceased North American Mycologists Including a Few European Mycologists," January 1975, unpublished and unpaged book held at the John N. Couch Biology Library, botany section, UNC–CH.

Jenkins, Charles F. "Asa Gray and His Quest for Shortia Galacifolia." *Arnoldia* 51.4 (1991): 4–11.

Lay, Lucy. "Artistic Development of the University Campus." *The Alumni Review, The University of North Carolina*. March 1926, 175, 177.

Linville, Susanne Gay. *Jennie: Jennie Coker Gay*. Privately printed, 1983.

Lobin, Wolgram, Stefan Giefer, Robert W. Krapp, Thomas Pauls, und Wilhelm Barthlott. *Bäume und Sträucher im Botanischen Garten der Universität Bonn*. Bonn: Botanischer Garten, 1998.

McVaugh, Rogers, Michael R. McVaugh, and Mary Ayres. *Chapel Hill and Elisha Mitchell the Botanist*. Chapel Hill, N.C.: The Botanical Garden Foundation, 1996.

Madry, R. W. "Beauty of Chapel Hill is Monument to Coker." *News and Observer* [Raleigh, N.C.] 26 Aug. 1944, 6.

Meriwether, Margaret Babcock, ed., app. W. C. Coker. *The Carolinian Florist of Governor John Drayton of South Carolina, 1766–1822, with Water-color Illustrations from the Author's Original Manuscript and an Autobiographical Introduction*. Columbia, S.C.: South Caroliniana Library, USC, 1943.

Methven, Andrew S. "William Chambers Coker," *McIlvainea, Journal of American Amateur Mycology* 14.2 (2000): 27–33.

*North Carolina Botanical Garden, University of North Carolina at Chapel Hill: Master Plan, A Guide for Development*. n.p., March 1992.

Pilkey, Orrin H., William J. Neal, Stanley R. Riggs, Craig A. Webb, David M. Bush, Deborah F. Pilkey, Jane Bullock, and Brian A. Cowan. *The North Carolina Shore and Its Barrier Islands: Restless Ribbons of Sand*. Durham: Duke UP, 1998.

Powell, William S. *The First State University: A Pictorial History of the University of North Carolina*, Bicentennial Edition. Chapel Hill: U of North Carolina P, 1992.

Radford, Albert, Harry E. Ahles, and C. Ritchie Bell. *Manual of the Vascular Flora of the Carolinas*. 2nd ed. Chapel Hill: U of North Carolina P, 1968.

Radford, Laurie Stewart. "The History of the Herbarium of the University of North Carolina at Chapel Hill, N.C." Unpublished essay, copyright applied for, 1998.

*Recollections of the Major: James Lide Coker 1837–1918*. Hartsville, S.C.: Hartsville Museum, 1997.

Ricklefs, E. and Dolph Schluter. *Species Diversity in Ecological Communities: Historical and Geographical Perspectives*. Chicago: U of Chicago P, 1993.

Rogers, Donald P. *A Brief History of Mycology in North America*. Reprint in augmented form, Mycological Society of America. Cambridge: Harvard University Printing Office, 1981.

Rogers, James A. with Larry E. Nelson. *Mr. D. R.: A Biography of David R. Coker*. Hartsville, S.C.: Coker College P, 1994.

Sanders, Albert E., and William D. Anderson Jr. *Natural History Investigations in South Carolina from Colonial Times to the Present*. Columbia: U of South Carolina P, 1999.

Sargent, Ralph M. *Biology in the Blue Ridge: Fifty Years of the Highlands Biological Station 1927–1977*. Highlands, N.C.: The Highlands Biological Foundation, Inc., 1977.

Savage, Henry, Jr., and Elizabeth J. Savage. *André and François André Michaux*. Charlottesville: UP of Virginia, 1986.

Shaffner, Randolph P. *Heart of the Blue Ridge: Highlands, North Carolina*. Highlands: Faraway Publishing, 2001.

Simpson, George Lee, Jr. *The Cokers of Carolina, A Social Biography of a Family*. Chapel Hill: U of North Carolina P, 1956.

Snider, William D. *Light on the Hill: A History of the University of North Carolina at Chapel Hill*. Chapel Hill: U of North Carolina P, 1992.

Stucky, Jon M., and Milo Pyne. "A New Species of Liatris (Asteraceae) from the Carolina Sandhills." *Sida* 14.2 (1990): 189–208.

Totten, H. R., and J. N. Couch. "William Chambers Coker." *Journal of the Elisha Mitchell Scientific Society* 70 (December, 1954): 116–18.

Troyer, James R. *Nature's Champion: B. W. Wells, Tar Heel Ecologist*. Chapel Hill: U. of North Carolina P, 1993.

Walser, Richard. *Thomas Wolfe Undergraduate*. Durham, N.C.: Duke UP, 1977.

Wells, B. W. *The Natural Gardens of North Carolina*. 1932. Chapel Hill: U of North Carolina P, 1967.

Wilson, Louis R. *The University of North Carolina, 1900–1930: The Making of a Modern University*. Chapel Hill: U of North Carolina P, 1929.

Wolfe, Thomas. *The Letters of Thomas Wolfe, Collected and Edited with an Introduction and Explanatory Text by Elizabeth Newell*. New York: Charles Scribner's Sons, 1956.

———. *Look Homeward, Angel*. New York: Charles Scribner's Sons, 1957.

———. *Passage to England: A Selection*. Ed. Suzanne Stutman and John L. Idol. Rocky Mount, N.C.: Walker Printing, The Thomas Wolfe Society, 1998.

———. *The Web and the Rock*. Garden City, N.Y.: Sundial Press, 1940.

# Index

Wolfe, Thomas, 77, 92–94
Woollen, C.T., 95, 96
Wright, Henry, 41
Wyatt, Robert, 73, 74

*Xanthoxylum clàva-hérculis*, 55, 56

Yucca
  in W.C. Coker's garden, 115

*Zenobia cassinefolia*, 126
Ziegler, Arthur William, 131